THE COSMOS on a SHOESTRING

Small Spacecraft for Space and Earth Science

Liam Sarsfield

Prepared for the Office of Science and Technology Policy

Critical Technologies Institute
RAND

In the wake of Federal budget reductions, both the civil and military space programs have turned to small spacecraft to meet mission requirements at lower costs. A RAND study for the Office of Science and Technology Policy (OSTP) and the Office of Management and Budget (OMB) has investigated smaller programs and the role they play in meeting important national objectives in space. This report reviews the factors leading to the growth in small-spacecraft programs and the effects of associated cost-reduction approaches. Although the report includes references to some Department of Defense (DoD) missions, it focuses on National Aeronautics and Space Administration's (NASA's) science missions, and the conclusion and recommendations presented reflect this concentration.

As small spacecraft take on a more central role, it is critical that decisionmakers appreciate the dynamics of smaller programs and maintain realistic expectations of their potentials. As part of this study, 12 small spacecraft programs were analyzed. The purpose of this analysis was to examine spacecraft development trends to highlight areas in which new strategies have proven effective in reducing cost and increasing performance. At the request of OSTP and OMB, special attention was focused on the role that technology plays in small spacecraft missions and the processes used to evaluate the performance of these missions.

The insights presented in this report are related to many elements of current civil and military space policy and should be of interest to OSTP and OMB officials in oversight and policy positions. Additionally, it is hoped that the report will serve as a useful reference for DoD and NASA program officials charged with streamlining future missions. Some aspects of the discussion related to technology programs might also be of interest to NASA managers responsible for the new systems needed to meet future scientific requirements. Finally, it is hoped that the analysis could assist officials challenged with creating metrics for Federal research and development programs.

This study was conducted as part of the Space Studies program of the Critical Technologies Institute (CTI) and was sponsored by the Office of Space Science, NASA. CTI is a federally funded research and development center (FFRDC) for studies and analysis, established by OSTP.

CONTENTS

FIGURES

TABLES

Small spacecraft have evolved to become a key element of the civil space program. In the wake of Federal budget reductions, civil and military space programs have had to downsize. Small spacecraft are an important means of maintaining scientifically viable programs within these tighter budgets. Within NASA, small programs also substantiate the "faster, better, cheaper" management and design strategies created in response to the National Performance Review.

The need to reduce mission cost is certainly a prime driver in the shift to small, scientific spacecraft, but there are other factors:

- Large missions were often taking too long to complete and were often considered unresponsive to the needs of the earth and space-science communities.[1]

- New miniature technologies have enabled smaller, more capable spacecraft to be constructed.

- The loss of large, expensive spacecraft, such as the Mars Observer, prompted a desire to spread mission risks.

It is also important to note that NASA's shift to small spacecraft is, more correctly, a return to a design philosophy largely abandoned as spacecraft size and complexity grew in proportion to expanding science requirements and the lift capacity of new, larger boosters. NASA's new small spacecraft rely on advanced technology more than their predecessors did, allowing them, in many cases, to return a surprising amount of scientific data. Typically, these missions take under three years to develop and cost, on average, $145 million.[2]

---

[1]Small spacecraft began to play a renewed role in NASA's mission portfolio before the agency's space-science budget began to decline in Fiscal Year 1992 (FY92). Their re-emergence was largely based on demands from the science community for NASA to build spacecraft more quickly.

[2]Based on data compiled during this study.

As the recent Mars Pathfinder mission aptly demonstrates, small spacecraft are capable explorers, able to pursue important scientific objectives while still capturing the imagination of the American public.

OBJECTIVES OF THE STUDY

Small spacecraft will likely play an increasingly important role in both the military, civil, and commercial space programs. One-quarter of NASA's current investment in space and earth science is spent on small programs, an amount certain to increase as larger programs, now under development, conclude. Within DoD, the Air Force, the Navy, and the National Reconnaissance Office are exploring ways to shift assets to smaller platforms that can be deployed more rapidly at lower cost.

As small spacecraft play a more central role in national space policy, it is important that decisionmakers understand the dynamics of smaller programs and maintain realistic expectations of their potentials. Developments related to small programs also offer new options in terms of policy formulation and implementation. With these thoughts in mind, OSTP and OMB asked RAND to undertake a study of small programs with three objectives:

1. To inform policymakers regarding the shift to small spacecraft within the civil space program. Specifically, OSTP and OMB requested that RAND address four questions:

    - What roles are small spacecraft currently playing in the civil space program?

    - What strategies have proven especially effective in reducing cost and increasing performance of small spacecraft?

    - What role does advanced technology play in the process of building small spacecraft?

    - How should government evaluate civil small spacecraft programs to ensure that objectives are met cost effectively?

2. To identify issues related to NASA's increased reliance on small spacecraft.

3. To provide policy and program-level recommendations based on the research conducted during the study.

Examining small spacecraft programs required a multidisciplinary approach and the study of many dimensions of NASA's programs. RAND's methodology was to rely heavily on close interactions with the NASA offices responsible for conducting space research and the many supporting offices that develop technology and assist in the process of building and testing spacecraft. There were many visits to contractor facilities to review development practices. To help gauge NASA's currently methods, a set of representative small spacecraft missions was selected. Technical data were secured directly from the program office responsible for each of the missions studied. NASA Headquarters' Office of the Chief Financial Officer supplied the cost data. Phone surveys helped to answer specific technical and cost questions. At the midpoint of the study, a workshop was conducted at the RAND offices in Washington, to which engineers and managers from NASA, DoD, industry, and academia were invited. **Because the study's scope was extensive, a large portion of the analysis and many of the conclusions and recommendations are relevant to NASA's programs in the broadest sense.**

STUDY FINDINGS—UNDERSTANDING THE SHIFT TO SMALLER SPACECRAFT

Based on the initial questions from OSTP and OMB, the study undertook a comprehensive review of practices related to the management and engineering of small spacecraft. While these findings are drawn from a review of small civil spacecraft, they may provide useful insights into similar programs being developed in the military and commercial space sectors.

The Role of Small Spacecraft

NASA's current generation of small spacecraft is capable of impressive levels of performance. Small spacecraft fulfill important roles in earth science, astrophysics, space physics, and planetary science. Yet, despite performance improvements, they cannot, and were not intended to, produce a science program equal in content to past programs. Rather than replace their larger counterparts, small spacecraft currently exploit opportunities that have been identified by previous missions, perform focused investigations, and serve in a precursor role.

There is no single class of small spacecraft. They range from relatively simple spacecraft, represented by those built by university students and the amateur radio satellite community, to spacecraft that rival the complexity of their larger counterparts. Clearly, these smaller spacecraft can be built in less time than it takes to build a large one. Less material and engineering time are required, launch systems are smaller and less costly, and they cost less to operate. In an absolute sense, small spacecraft are "cheaper." In a relative sense, however, they are generally more expensive for three reasons:

- To meet the objectives of science, they remain complex, often costing appreciably more per kilogram than larger spacecraft.[3]

- There is an economy of scale in the launch market—it costs more per kilogram to launch smaller spacecraft.

- A greater degree of risk has been tolerated in the development and operation of small spacecraft, so invested funds are exposed to a higher potential for failure.

Small spacecraft are, therefore, currently premium items in the inventory of space science missions. The small spacecraft economy could change substantially, however, if performance continues to increase while development costs drop. This underscores the importance of continuous refinement in spacecraft design and development practices and advanced, high-performance technology. Both developments allow spacecraft to be built less expensively while achieving higher levels of performance. It is indeed possible that the capabilities of small spacecraft could improve to the point where they support the bulk of future mission requirements, making them highly cost-effective instruments of exploration.

Cost and Performance of Small Spacecraft

NASA has been challenged with crafting a program that continues to produce meaningful science within the constraints of the available budget. To accomplish this goal, the agency has significantly reordered internal priorities, focusing on mission performance and cost. This change in focus can be viewed as a shift from scientific excellence alone to a broader objective: *mission excellence.* Previously, the requirements of science had been the overwhelming driver in mission design. Today, science, cost, and technical requirements carry more or less equal weight. This shift was necessitated by the fact that simply limiting the size of missions, in response to tighter budgets, would not ensure a scientifically viable program. NASA has been driven to build highly capable, smaller spacecraft and, in the process of doing so, to improve its own efficiency. This broader approach to mission excellence has reduced the cost of spacecraft missions, of any size, by approximately 20 percent.

To achieve lower cost and better performance, NASA has relied more heavily on a maturing commercial sector for the building of spacecraft and on the science community for the management of missions. Over the past 39 years, NASA;

---

[3]Small planetary spacecraft are an important exception—recent missions are costing less per kilogram than their predecessors. NASA also sponsors a limited number of simple, inexpensive spacecraft built by university researchers.

DoD; the private sector; and a host of foreign, university, and amateur builders have built an extraordinary variety of spacecraft. As a result, the skills and resources required to build a spacecraft have proliferated widely. What was an experimental activity has matured into a market economy for space systems. Today, for example, a commodity market in high-performance systems and components, such as Global Positioning System (GPS)–based guidance packages and mass-memory devices, provides engineers with many more opportunities to procure equipment instead of building custom components. Significant, too, is the proliferation of small communication satellites, which expands the available inventory of components and subsystems, many of which meet the requirements of science spacecraft. The marketplace includes entire spacecraft buses (the portion of the spacecraft that provides power, communications, and other operating resources) to which instruments can be added to complete a mission at considerably lower cost than a custom design. Not all of NASA's missions can pursue such an option, and commercial systems are not always the most cost-effective solution, but an expanding supply of commercial equipment is a clear trend in the development of space systems. The growing commercial marketplace also brings with it mass-production product reliability.

NASA's management roles are also changing. The agency has revitalized and streamlined its Announcement of Opportunity (AO) process, which caps mission costs and invites development of spacecraft under fixed-price or performance-based contracts. Management of NASA's science missions is increasingly handled by a competitive selection of a Principal Investigator (PI). Under PI-mode management, the investigator is responsible for planning and implementing the mission and for delivering scientific results.

The Role of Advanced Technology

While size does not imply a demand for new systems, there is a close relationship between small spacecraft and technology programs. Shorter development timelines allow small spacecraft to approach the state of the art more closely than was possible on larger missions that took years to develop. Spacecraft builders have also perceived a higher tolerance of risk on small missions (which, in fact, may be discordant with national policy interests) and have approached nontraditionally the use of new designs. Most importantly, however, new technology is necessary to meet science requirements that continue to be ambitious.

The need to integrate new, high-performance technology presents the program managers with a dilemma. New technologies often carry with them significant cost and schedule risks, making them difficult to manage within the framework of small programs operating with very little reserve. The need to mature technology more rapidly and more thoroughly has led NASA to (1) amplify efforts to plan and coordinate technology programs and (2) rely more heavily on missions dedicated to demonstrating new technologies in orbit.

NASA spends approximately 2 percent of its annual budget on spacecraft technology programs. The need to maximize returns from this modest investment underscores the importance of NASA's technology planning process. Past reviews of NASA's technology plans have criticized agency practices. In response, the agency produced the 1992 Integrated Technology Plan (ITP). While the ITP attempted to coordinate technology projects centrally, many elements of a plan were missing: a clear methodology for managing technology development, a clear tie between technology projects and future mission requirements, and a means of evaluating success and failure. To elevate the importance of technology planning, NASA disbanded the Office of Space and Advanced Technology (OSAT) in 1996, placing responsibility for the planning of all spacecraft technology programs under a new Office of Technology (OT) within the Office of the Administrator. A central objective of this realignment was to refine and reissue the ITP.

Many factors complicate the task of preparing an integrated plan. Perhaps the most important challenge NASA faces is the fact that the agency has traditionally had a dual mandate: to conduct science *and* develop technology. This duality has spawned separate cultures within the agency. *Mission-oriented* groups are closely aligned with, and responsive to, in-house spacecraft builders. These groups supply important incremental advances. *Research-oriented* groups have not been associated directly with flight programs unless a specific technology was being tested. Yet their basic research orientation has supplied some of the more revolutionary advances in spacecraft systems.

Beyond the planning of new technology is the need to reduce the cost, schedule, and technical risks of using a new design. The performance of a new design can usually be adequately evaluated using low-cost, ground-based approaches. Occasionally, however, a new technology must be tested in space, and several low-cost test methods are available. These methods have proven successful, and *new technology has traditionally not been the source of mission failure.* NASA does exploit low-cost options. Today, however, the agency relies heavily on dedicated technology-demonstrator spacecraft. The cost of these demonstrators, such as the New Millennium series, can exceed the cost of small science missions. In FY96, nearly one-third of NASA's annual $1 billion investment in small missions was used to construct technology demonstrator spacecraft. This level of investment reflects a belief within the agency that *revolutionary* technology is urgently needed to establish small spacecraft as mainline research platforms.

Performance Measurement

The key issue in terms of evaluating small spacecraft programs is formulating a response to the requirements of the 1993 Government Performance and Results Act (GPRA). Agencies must prepare both strategic and performance plans to comply with GPRA. NASA's Mission to Planet Earth and Space Science enterprises both maintain detailed strategic plans; performance plans are, however, more challenging. Spacecraft programs have components that are basic research, applied research, technology development, and production engineering, each with elements that are quantifiable and nonquantifiable. This suggests a hybrid performance plan in response to GPRA requirements.

NASA has taken significant steps toward improved accountability and performance measurement. The various small spacecraft programs evaluated in this study were, for example, readily able to provide mission cost data with a high degree of confidence. New full-cost accounting methods will likely further increase the accuracy of budget reporting. In terms of technical performance, the study found that NASA has an assortment of useful measures with which to report programmatic and scientific performance. Assessment of risk, however, is missing from the field of performance measurement.

Program evaluations are most valid when they include measurements of performance, cost, *and* risk. Commercial firms assess progress by measuring production cost, performance outputs, and product quality or reliability. The symmetric measurement of these parameters is very important, since cost or performance improvements could be offset by decreased reliability. While NASA has the means to measure cost and performance with accuracy, an equivalent metric for risk is needed.

The shift to small spacecraft within NASA brings with it a change in how risk is treated. Past missions practiced *risk avoidance* and expended a significant portion of mission funds eliminating potential sources of failure. Today's smaller missions cannot afford elaborate risk avoidance practices and have instead shifted to a *managed risk* approach.[4] While minimized to the greatest extent practical, the risks associated with the current generation of small spacecraft are generally acknowledged to be higher. A significant source of this increased risk is associated with launch, since small launchers have yet to demonstrate levels of reliability achieved by larger, mature systems.

---

[4]Managed risk means treating failure probabilities simply as another engineering variable in the process of building spacecraft. It marks a departure from earlier risk avoidance strategies, in which spending and level of risk reduction were often not correlated.

Risk is not usually regarded as a performance measurement variable and is not easy to calculate. Shifting to a managed risk approach, however, requires that NASA place a greater emphasis on risk measurement. The Office of Safety and Mission Assurance (OSMA) has been refining quantitative methods of measuring risk that are easier to use. Although geared to supporting the engineering function, these measurement techniques could have broader application.

An improved ability to calculate risk supports efforts to measure performance, while also providing a mechanism for communicating risk. It is, therefore, possible for policymakers to contemplate using risk reduction as a goal for the civil space program.

The refinement of risk measurement techniques would likely take some time. In the interim, NASA's measurement of spacecraft reliability could serve as an important measure of technical capability. The reliability of scientific spacecraft has improved steadily. Failures, when they do occur, are generally less severe, and spacecraft are living well beyond initial design points. This is an achievement for which NASA does not now receive recognition. Several long-term technical trends indicate that the reliability of spacecraft could increase considerably further:

- Knowledge of failure mechanisms is expanding rapidly.

- Procedures used to test space systems are more accurate.

- Component and subsystem reliability is rapidly improving.

- Design processes are more robust.

The potential for significant improvements in the reliability of space systems reinforces the notion that it could be useful as a performance indicator.

KEY ISSUES

The research conducted during this study identified several issues that are relevant not only to the continued development of small spacecraft but also to improving the performance of NASA's missions. Some of these issues relate directly to matters of national policy. Most of these issues, however, related to how missions are managed and evaluated within NASA. The discussion that follows summarizes these issues, then provides recommendations that could help deal with these issues.

National Policy

Profound change has swept through NASA programs in recent years. While the changes were designed to improve the ability of the agency to conduct cost-effective science investigations, the study noted several emerging issues.

In terms of budget, there are some indications that NASA's small spacecraft are operating at a limit in terms of supporting the requirements of the science community. NASA's scientific advisory boards have expressed concern that further erosion of investments in science missions would threaten the ability of programs to retain world-class researchers. Of particular concern are the planetary sciences, in which funding constraints have significantly reduced mission capabilities. Future Mars missions, for example, may not be sufficiently funded to carry adequate instrumentation. RAND workshop participants underscored the importance of allowing the science community to specify *appropriately sized programs*, optimized to return the greatest amount of data that can be generated within the available budget.

Another issue of importance to national space policy, related to the shift to smaller spacecraft, is the agency's changing role as a builder of space systems. Following the directives of the National Space Policy, NASA is mandated to develop spacecraft only when the unique technical capabilities of a field center are required. As the space marketplace has matured, NASA has enjoyed more-frequent opportunities to turn to the private and university sectors to build spacecraft. Under PI-mode management, NASA is also outsourcing the management function. Increasingly, NASA is no longer a builder or manager of spacecraft programs.

NASA scientists and spacecraft-development teams must compete in this environment. It could prove difficult for government spacecraft teams to compete, however, as full-cost accounting procedures take effect. The shift to full-cost accounting could raise the historical cost of building a spacecraft in house by as much as 40 percent.

Streamlined AOs and the reliance on PI-mode management have helped NASA respond to Federal budget reductions, but the agency's core competencies could erode as a result. NASA operates extensive spacecraft design and test facilities and is an important source of training for young engineers and managers. Staff reductions reflect the reduced requirements of smaller missions, but the magnitude of change suggests that functional realignments within the agency might be needed. The desire of policymakers to reduce costs and ensure program effectiveness should be balanced against a need to redefine, protect, and strengthen NASA's core competencies and the pursuit of mission excellence.

A related issue is preserving program integrity in small spacecraft programs that are managed at the lean limit. Efforts to reduce spending and improve the performance of small spacecraft programs have proven effective. There are, however, indications that program integrity is being threatened by lean operating margins. Some small spacecraft builders feel they must overstress development teams to meet aggressive cost and schedule targets. In some cases, this has resulted in the loss of key personnel, a significant factor for smaller teams. The risk of losing valuable managers and engineers for the limited profits associated with building a small spacecraft could also cause private firms to forgo competing on future development contracts. Tight schedules and lean operating budgets can also increase risk, although increased failure potentials are rarely calculated. In many cases, resources are also insufficient to document lessons learned, to train new personnel adequately, or to invest in new capital equipment. Options to further streamline programs should not overlook the long-term value of maintaining a strong and capable supporting infrastructure.

Technology Planning and Implementation

NASA has placed a great deal of attention on accelerating the development of the advanced systems needed to improve the performance of small spacecraft. However, two important issues related to NASA's space technology program remain: (1) the effectiveness of the ITP and (2) the cost-benefit ratio of the investment being made in technology demonstrator missions.

Creating the ITP requires NASA to conjoin several other technology planning efforts. The authority of the ITP over these various programs is not clear. Technology planning occurs within several program offices and at various field center locations. To be fully integrated, the ITP must assimilate or coordinate these efforts, in turn implying a high degree of trust between those building spacecraft and those developing technology. NASA has created an internal contract structure to ensure that technology, planned within the framework of the ITP, will be ready when needed. For such a contract to work, however, an assurance is needed that the techniques and terminology used to measure and describe the readiness of a new design are common to both technologists and the spacecraft builder. Here, the ITP could potentially rely on the experience of OSMA to provide this assurance.

The level of NASA's investment in technology demonstrator missions may be excessive in terms of potential returns and the agency's mission. Both NASA and DoD utilize demonstrator spacecraft, but they have different objectives regarding their use and different implementation strategies. NASA seeks to prepare technology for use on future single-purpose science missions. Military demonstrator missions are important precursors to operational networks, such as GPS and Milstar. These networks represent multiyear, multi–billion-dollar investments that support critical national security requirements. In general, military flight demonstrators are low-cost endeavors, attempting to test the capabilities and operation of a single technology. On the other hand, NASA's dedicated demonstrator spacecraft are considerably more ambitious undertakings. Significant cost and schedule risk is associated with these missions, since several new technologies are usually incorporated into the spacecraft.

NASA's demonstrator missions also seek to test new technology in actual mission environments, thus overcoming potential resistance from scientists concerning future use. Yet the space environment is being characterized with greater accuracy so that performance of a technology in a given environment can be predicted with higher confidence. It is also difficult to extrapolate experiences with a new technology from one environment (for example, the environment in near-Earth orbit) to another (the environment encountered by a deep-space mission).

Investments in dedicated technology-demonstrator missions should be balanced against both verified scientific need and gains that could accrue from expanding the base of fundamental research and technology programs. In establishing this balance, it is important that the risks these missions are designed to retire are real and not simply perceived. Requirements for these missions should be generated from within NASA's ITP, which, in turn, should apply risk assessments based upon actual flight failure and performance data.

Perhaps the best way to assess the effectiveness of both NASA's ITP and the role of demonstrator spacecraft would be a more thorough review process. Past reviews of NASA's space technology programs could be regarded as insular. Projects were reviewed, but the review process itself was largely internalized. The 1990 Augustine Committee recommended that technology funds be allocated based on a review by experts outside of NASA. RAND workshop participants were unanimous in their agreement with this recommendation. To support such a review, the ITP must contain cost, schedule, and performance data for the myriad of technology projects that NASA sponsors. Demonstrator missions would have to be closely tied to the requirements contained within the ITP, and data would need to be provided to support the necessary cost-benefit analyses.

Measuring and Reducing Risk

Since the inception of the space program, success in space has bolstered national prestige and international perceptions of U.S. technological prowess. Space missions, even small ones, also represent a significant investment of Federal resources. The risks associated with space mission are, therefore, a matter of national importance.

At issue is maintaining the pattern of success associated with NASA's space-science program. While the study found that, historically, science spacecraft have demonstrated increasing reliability, this trend might not continue with the current generation of small spacecraft.[5] A consequence of the shift to managed risk is a greater potential for failure. Less money is generally available to smaller programs to test spacecraft functions and operational procedures prior to launch. Small spacecraft are also generally less robust. Consequently, policymakers should anticipate that failures will occur more frequently than in the past.

This observation highlights the importance of NASA's research in the area of high-reliability systems. Efforts to reduce the potential for failure by applying more-reliable components, better testing, and advanced design techniques should receive greater attention.

Measuring Performance

As noted above, the risks associated with NASA's space-science missions are a matter of national importance. Yet the communication of risk between NASA and policymakers occurs only on rare occasion.[6]

The policymaker can no longer assume that risks are being minimized and should be aware of the level of risk reduction that is being achieved with available funds. NASA's new managed-risk strategy carries with it both a need and an opportunity to communicate risk information more effectively. The communication of risk between NASA and the policymaker could take a form similar to an investment portfolio. This would require benchmarking current programs and evaluating future progress.

RECOMMENDATIONS

Small spacecraft will remain an important element of the space program for the foreseeable future. The potential exists for a renaissance in our understanding of the earth and space within the confines of a limited budget. Based on the

---

[5]One notable exception in the trend toward higher reliability is the performance of mechanical systems. Mechanical devices are a source of significant failures and warrant special attention in the effort to improve performance.

[6]The health-and-safety aspects of launching a spacecraft carrying a radioactive power source, for example, require formal communication of risk, since launch approval is required from the Office of the President of the United States.

questions that framed the study, the recommendations can be divided into four categories: civil space-policy objectives, performance improvement, improving technology planning and implementation, and measuring performance.

Civil Space Policy Objectives

- Establish a goal within the National Space Policy calling for NASA to pursue mission excellence in the design and development of science spacecraft. This goal would formally recognize the important role that the agency plays in improving the performance of space systems, which, in turn, strengthens our leadership in both the military and commercial space sectors.

- Conduct a review of NASA roles and missions in relation to a mature commercial space sector. This review should seek to identify NASA's unique strengths and capabilities in the areas of technology development and spacecraft design, development, and operations. It should also clearly identify the functions that must be retained and reinforced in regard to agency core competencies.

Improving Technology Planning and Implementation

- Firmly establish the ITP as NASA's focal point for the coordination of all instrument, spacecraft, and ground-system technology initiatives. Merge current spacecraft technology-development programs under the umbrella of the ITP. Within the ITP, create guidelines that establish a balance between basic research and nearer-term development projects.

- As a supplement to the ITP, prepare an annual report for instrument and spacecraft research and development projects. This report should include budgets (past, current, and projected spending), milestone schedules, and performance benchmarks.

- Initiate requirements for future technology flight-demonstrator missions from within the framework of NASA's ITP. The definition of these missions should emerge from a process that validates that flight in space is the only method of adequately retiring the risk of using a new technology. Additionally, this process should validate that technology demonstration is being pursued by the most cost-effective means.

- Evaluate the use of incentive awards to spacecraft development teams for advanced technologies that can be matured, documented, and prepared for transfer to other spacecraft developments and/or terrestrial applications.

- Examine the potential for integrating the product assurance function into the technology planning and implementation process. The goal of this examination would be to evaluate whether product assurance engineers can assist technologists to prepare their products in a form most readily integrated by the end user, the instrument or spacecraft designer.

- Forge more cooperative alliances within the spacecraft-development community. Consider broader application of the partnership model the New Millennium Program created between science teams and developers of advanced technologies.

Risk Measurement and Reduction

- Increase funding for efforts to mature quantitative measurement of risk and reliability. New risk measurement techniques should be designed to support not only the technical management of missions but also the need for NASA program offices to communicate risks to the policymakers.

- Direct additional funds to research in high-reliability space systems and to the study of failure analysis, new test practices, and advanced design processes. Additionally, augment funding for test and evaluation of high-reliability mechanical systems for small spacecraft.

Measuring Performance

- Apply relative measurements of reliability within the earth- and space-science portfolios to monitor process improvement. Also, apply these measures (a) to communicate overall program risk between NASA and policy offices and (b) to distribute reserves within the programs.

- Formalize NASA's process improvement by benchmarking current spacecraft programs in terms of spacecraft cost, performance, and reliability and relate progress in terms of the change of these parameters within the earth- and space-science portfolios.

- Create a formal review process for the ITP. The review should involve senior technologists as peers. It should also include individuals who use space technology—mission scientists, who rely on technology to meet future requirements, and spacecraft designers, who must integrate new systems. The resulting user-peer review process should also involve external, unbiased agents, who can dispassionately assess the merits of the agency's progress on these programs. Peer review results should be reported annually in the ITP report supplement.

The author wishes to thank Jeff Hofgard of OSTP and Steven Isakowitz of OMB for their assistance and guidance in this research. Special thanks also go to Earle Huckins and Mary Kicza of NASA for their professional advice and support. The author would also like to thank two former OSTP staff members: Richard DalBello, who, as former Assistant Director for Space, helped formulate the initial research agenda, and Paul Regeon, whose technical knowledge helped illuminate many key study areas.

A central element of this study was obtaining open access to cost and technical details of the missions selected for review. The author wishes to thank the staffs of NASA's Discovery, Explorer, New Millennium, Small Spacecraft Technology Initiative, and Surveyor programs for their support. Jack Chapman was instrumental in obtaining details of the Clementine I mission. The author wishes to thank Gary Rawitscher of NASA's Office of the Chief Financial Officer for his patience and rapid response to repeated questions and inquiries. The support of Stuart Heller at the Jet Propulsion Laboratory (JPL) and Bernard Dixon and his staff at the Goddard Space Flight Center (GSFC) is also gratefully acknowledged.

The study benefited greatly from the kind assistance and technical insights of Orlando Figueroa of GSFC's Explorer Program and his Small Explorer team. The author wishes to expressly thank Charles Elachi, Kane Casani, and Barbara Wilson of JPL who, with their staff, took a great deal of time to provide briefings and references related to the New Millennium Program. Also at JPL, Carl Kukkonen of the Center for Space Microelectronics Technology, Michael Sander of the Space and Earth Science Programs Directorate, and David Smith and Nicholas Thomas of the Project Design and Architectural Support Office, provided valuable insights into technology trends and new spacecraft-design strategies.

A study of issues related to small spacecraft cuts across many technical disciplines and organizations, and the author would like to express appreciation to the many individuals in government, industry, and academia who took time to meet and discuss their various programs. A special note of appreciation goes to David Bearden of the Aerospace Corporation; Anthony Sabelhaus of TRW Spacecraft Operations East; Thomas Coughlin of the Johns Hopkins University Applied Physics Laboratory; William Grunenwald of Modern Technologies, Inc.; Michael Greenfield of NASA's Office of Safety and Mission Assurance; and Steven Cornford of JPL. The author also wishes to express his gratitude to those who participated in a two-day RAND workshop on small spacecraft development trends. The study's objectives were advanced greatly by the willingness of workshop participants to share their experiences openly.

The author is indebted to RAND colleagues Scott Pace, James Bonomo, and Bruce Don, who provided extensive and thorough reviews of the interim and final reports, and to James Dryden, for additional review and comments. Jay Falker also provided a thorough analysis of launch-vehicle cost and reliability.

The RAND support staff underpins the development of RAND studies and reports, and the author graciously thanks the many members who contributed to the workshop and final report. Special thanks to the RAND publication staff: Phyllis Gilmore, who tirelessly edited the final manuscript; Sandy Petitjean, who produced the many graphics and tables; and Alisha Pitts, who created the cover illustrations. Many thanks also to Julia Shaw and Amber Dumas for helping get the final product assembled, and to Gail Kouril for helping to assemble references and citations. The author, of course, remains responsible for the observations and judgments contained in this report.

GLOSSARY

ACS	Attitude control system
ADEOS	Advanced Earth Observation Satellite
AE-C	Atmospheric Explorer C
AEM-HCMM	Application Explorer Mission—Heat Capacity Mapping Mission
AM	Ante meridiem (the AM-1 is NASA's Earth Observing System satellite with a 10:30 a.m. descending node)
AMSAT	Radio Amateur Satellite Corporation
AO	Announcement of Opportunity
ARL	U.S. Army Research Laboratory
AT&MS	Advanced Technology and Mission Studies (Division of OSS)
ATD	Advanced Technology Development
AU	Astronomical unit
AXAF	Advanced X-Ray Astrophysics Facility
BMDO	Ballistic Missile Defense Organization
BOL	Beginning-of-life
C&DH	Command and data handling
CAD	Computer-aided design
CARMA	Computer-Aided Reliability and Maintainability Analysis (Texas Instruments model)
CATIA	Computer-Aided Three-Dimensional Interactive Approach (Boeing)
CATSAT	Cooperative Astrophysics and Technology Satellite
CDRL	Contract [data/document] requirements list
CER	Cost estimating relationship
CERES	Clouds and Earth Radiant Energy System
CGRO	Compton Gamma Ray Observatory
Chem	Earth Observing System Chemistry satellite
CMOS	Complementary metal-oxide silicon
COBE	Cosmic Background Explorer
CoF	Cost of facilities

COMPLEX	The Committee on Planetary and Lunar Exploration, of the Space Studies Board, NRC
CORBA	Common Object Request Broker Architecture
COTS	Commercial-off-the-shelf
CPU	Central processing unit
CR	Change request
CRAF	Comet Rendezvous and Flyby
CRUX	Cosmic Ray Upset Experiment
DARPA	Defense Advanced Research Projects Agency
DBOF	Defense Business Operating Rules
DoD	Department of Defense
DPM	Defects per million
DS	Deep space
ECN	Engineering change notice
EDAC	Error detection and correction
EEE	Electrical, electronic, and electromechanical
EIA	Electronic Industries Association
ELV	Expendable launch vehicle
EO	Earth observing
EPS	Electrical power system
ESS	Environmental-stress screening
ESSP	Earth Systems Science Pathfinder
EUVE	Extreme Ultraviolet Explorer
FAR	Federal Acquisition Regulation
FASA	Federal Acquisition and Streamlining Act
FAST	Fast Auroral Snapshot Explorer
FIT	Failures in time
FST	Flight System Testbed (JPL)
FTB-IMO	Flight Test Bed for Innovative Mission Operations
FUSE	Far Ultraviolet Spectroscopic Explorer
GAO	General Accounting Office
GAS	Get-Away Special; a Space Shuttle–based small payload carrier
GEO	Geosynchronous orbit
GFE	Government furnished equipment
GP-B	Gravity Probe B
GPRA	Government Performance and Results Act, passed in 1993 as Public Law #103-62
GPS	Global Positioning System
GSE	Ground support equipment
GSFC	NASA Goddard Space Flight Center
HETE	High Energy Transient Experiment
HRLL	High Reliability/Long Life (JPL systems initiative)

HST	Hubble Space Telescope
HTML	HyperText Markup Language
I&T	Integration and test
IAE	Inflatable Antenna Experiment
IC	Integrated circuit
ICD	Interface control document
ICE	International Cometary Explorer
IDIQ	Indefinite design, indefinite quantity
IEEE	Institute of Electrical and Electronics Engineers
IFMP	Integrated Financial Management Plan
IMAGE	Imager for Magnetopause-to-Aurora Global Exploration
IMDC	Integrated Mission Design Center (GSFC)
IPDT	Integrated Product Development Team
IPT	Integrated Product Team
IR	Infrared
IRAS	Infrared Astronomical Satellite
ISHM	International Society for Hybrid Microelectronics
ITP	Integrated Technology Plan
IUE	International Ultraviolet Explorer
JPL	NASA Jet Propulsion Laboratory
KGD	Known good die
LEO	Low earth orbit
LeRC	NASA Lewis Research Center
LMLV	Lockheed Martin Launch Vehicle
LP	Lunar Prospector
LV	Launch vehicle
M&S	Modeling and simulation
MA	Mission assurance
MAP	Microwave Anisotropy Probe
MCM	Multichip module
MGS	Mars Global Surveyor
MIDEX	Mid-sized Explorers
milspec	Military specification
milstd	Military standard
MO&DA	Mission operations and data analysis
MP	Mars Pathfinder
MS	Mars Surveyor
MSTI	Miniature Sensor Technology Integration
MTBF	Mean time between failures
MTPE	Mission to Planet Earth
NAFCOM	NASA/AF Cost Model
NASA	National Aeronautics and Space Administration

NEAR	Near Earth Asteroid Rendezvous
NMI	NASA Management Instruction
NMP	New Millennium Program
NRA	NASA Research Announcement
NRC	National Research Council
OAST	Office of Aeronautics and Space Technology
ODAP	Orbital Data Acquisition Program (Air Force)
OMB	Office of Management and Budget
OSAT	Office of Space Access and Technology
OSMA	Office of Safety and Mission Assurance (NASA)
OSS	Office of Space Science (NASA)
OSSA	Office of Space Science and Applications
OSTP	Office of Science and Technology Policy (White House)
OT	Office of Technology (NASA)
PA	Product assurance
PA&E	Program Analysis and Evaluation
PAM	Payload Assist Module
PDC	Project Design Center (JPL)
PEM	Plastic-encapsulated microcircuits
PFAD	Payload Flight Anomaly Database (JPL)
PFP	NASA's Program Financial Plan
PI	Principal investigator
PIDDP	Planetary Instrument Definition and Development Program
PM	Post meridiem
PSR	Program status report
R&A	Research and analysis
R&D	Research and development
R&PM	Research and program management
R&QA	Reliability and quality assurance
RAC	Reliability Analysis Center (Rome Air Force Base)
RADCAL	Radar Calibration
RaDiUS	Research and Development in the United States
RAO	Resource Analysis Office (GSFC)
RELTECH	Reliable Technology
REX	Radiation Experiment
RF Comm	Radio frequency communication
RFP	Request for proposals
RIF	Reduction in force
RTG	Radioactive thermal generator
S/C	Spacecraft
SAC	Scientific Applications Satellite (Argentina)
SAMPEX	Solar, Anomalous and Magnetic Particle Explorer

SBD	Simulation Based Design
SDIO	Strategic Defense Initiative Office
SEB	Single-event burnout
SEL	Single-even latchup
SELL	Space Engineering Lessons Learned (SELL) database (GSFC)
SESAC	Space and Earth Science Advisory Committee
SEU	Single-event upset
SIRTF	Space Infrared Telescope Facility
SMALLSAT	Small satellite
SMEX	Small Explorer
SNOE	Student Nitric Oxide Experiment
SOA	State of the art
SOAR	Spacecraft Orbital Anomaly Report (GSFC)
SP	Solar Probe
SR&T	Supporting research and technology
SRAM	Static random access memory
SSB	Space Studies Board
SSED	Space Systems Engineering Database
SSR	Solid-state recorder
SSTI	Small Spacecraft Technology Initiative
STC	Spacecraft Technology Center (Martin Marietta)
STDB	Space Technology Data Base
STEDI	Student Explorer Demonstration Initiative
STEP	Space Test Experiment Program
STP	Space test program
STS	Space Transportation System
SWAS	Submillimeter-Wave Astronomy Satellite
TDRS	Tracking and Data Relay Satellite
TERRIERS	Tomographic Experiment using Radiative Recombinative Ionospheric EUV (Extreme Ultraviolet) and Radio Source
TiPS	Tether Physics and Survivability Experiment
TMC	Total mission cost
TOMS	Total Ozone Mapping Spectrometer
TRACE	Transition Region and Coronal Explorer
TRL	Technology readiness level
TRMM	Tropical Rainfall Mapping Mission
UNEX	University Explorer
USAF	U.S. Air Force
VRML	Virtual Reality Modeling Language
WBS	Work Breakdown Structure
WIRE	Wide-Field Infrared Explorer

WWW	World Wide Web
XTE	X-Ray Timing Explorer
ZBR	Zero-base review

Chapter One

INTRODUCTION

In an era of limited resources, government is depending more heavily on small spacecraft to attain important civil and military space goals. Each of the National Aeronautics and Space Administration's (NASA's) science disciplines has at least one program dedicated to small spacecraft development. The agency's technology programs are also shifting to initiatives to develop systems and subsystems geared to smaller spacecraft. The Air Force, Navy, and the National Reconnaissance Office are all exploring ways to shift assets to smaller platforms that can be deployed more rapidly at lower cost.

A Department of Defense (DoD) decision to meet operational needs with smaller spacecraft will be based on the military's 30-year reliance on small spacecraft for advanced technology demonstration. NASA's shift to small spacecraft, on the other hand, is actually a return to a design philosophy largely abandoned as spacecraft size and complexity grew in proportion to an expanding set of science requirements.

The small spacecraft phenomenon is attributed to shrinking Federal budgets and broad-based efforts to streamline government programs. Within NASA, small programs are considered hallmarks of a "faster, better, cheaper" philosophy, substantiation that more can indeed be done with less. In the course of this research, however, it became clear that other factors have also driven the shift to smaller platforms. New miniature technologies have, for example, enabled construction of smaller spacecraft. Large missions were often taking too long to complete and proving unresponsive to the needs of the space-science community. Major losses, such as the Mars Observer spacecraft, have also caused program managers to reevaluate risk and the wisdom of relying on single, large spacecraft. These and other factors are important contributors to the emergence of small spacecraft.

This report provides a review of current trends in small spacecraft development and analyzes the effects of new strategies aimed at reducing cost and increasing spacecraft performance. This report also examines the role of technology and the process by which advanced systems are matured for use in small spacecraft. It suggests some reasonable expectations for future small spacecraft missions and provides thoughts for improving the way spacecraft development and technology programs can be better linked.

SMALL SPACECRAFT DEFINED

There is no official definition of a small satellite. Various studies have set different mass definitions. A recent textbook (Wertz and Larson, 1996) defines a small satellite as having a dry mass under 400 kg. The Center for Satellite Engineering Research at the University of Surrey, England defines a "mini" satellite as being between 100 and 500 kg. Other terms have been used to describe small spacecraft, such as *Lightsat* and *Cheapsat.* The term of choice depends largely upon the perspective of the developer. The Radio Amateur Satellite Corporation (AMSAT) might, for example, consider a 500-kg spacecraft huge.

For the purpose of this study, small spacecraft have been defined as those with a dry mass of less than approximately 500 kg. Using mass as a descriptor is somewhat misleading in that it fails to distinguish spacecraft. It is possible, for example, to conceive of a program with a small budget that produces a heavy spacecraft. Conversely, one could imagine a relatively lightweight spacecraft with a high development cost.[1] In practice, however, the 500-kg definition does a good job of focusing the study on programs that have pursued low-cost options, in terms of spacecraft development, management, and operations. In the body of this report, therefore the term small spacecraft is sometimes used interchangeably with small programs.

BACKGROUND

The first spacecraft launched into space were small. The Pioneer and early Explorer satellites were simple, inexpensive designs built to answer basic questions about the earth and near space. As knowledge and booster capability expanded, so too did the size and complexity of spacecraft. In the 70s and 80s, major research spacecraft grew to cost more than $1 billion and take nearly a decade to develop.[2]

---

[1]NASA's Gravity Probe B (GP-B) is an example of a spacecraft with a development cost disproportionate to its weight. GP-B is pressing beyond the state of the art in such areas as precision gyroscopes and cryogenic systems. Projected to cost in excess of $2.6 billion, the GP-B spacecraft is three times as expensive, in terms of dollars per kilogram, than the small spacecraft reviewed in this study.

[2]Some examples are the Compton Gamma Ray Observatory (CGRO), which took 9 years and cost $0.8 billion, Viking (7 years and $1.2 billion), and the Hubble Space Telescope (HST) (13 years and $1.5 billion), according to NASA budget data.

Today, a confluence of technical, political, and economic factors urges smaller, more focused spacecraft programs. The new small spacecraft are, however, a far cry from the pathfinder missions of the late 50s and early 60s. These new missions rely on advanced technology (higher performance systems and components that have not previously flown in space) far more than their predecessors. In some cases, advanced systems are allowing small spacecraft to return a surprising amount of scientific data. The time required to develop science spacecraft has also dropped to approximately three years, with an average mission cost of $145 million.[3] On the horizon, armadas of small spacecraft could create a virtual human presence in the solar system.[4]

Commercial firms are also proliferating small communication satellites in low earth orbit (LEO) to provide unprecedented global network services.[5] Government spacecraft programs are expected to benefit from the large private-sector investment in new systems, components, and production techniques.

STUDY PURPOSE

As small spacecraft play a more central role in national space policy, it is important that decisionmakers understand the dynamics of smaller programs and maintain realistic expectations of their potentials. Developments related to small programs also offer new options in terms of policy formulation and implementation. With these thoughts in mind, the Office of Science and Technology Policy (OSTP) and the Office of Management and Budget (OMB) asked RAND to undertake a study of small programs with three objectives:

1. To inform policymakers regarding the shift to small spacecraft within the civil space program. Specifically, OSTP and OMB requested that RAND address four questions:

   - What roles are small spacecraft currently playing in the civil space program?

   - What strategies have proven especially effective in reducing cost and increasing performance of small spacecraft?

---

[3]These averages are based on an analysis of 12 NASA space-science spacecraft in the New Millennium, Discovery, Surveyor, and Explorer series. Data were provided to RAND by the respective NASA program offices. See Appendix A.

[4]Remarks of the President's Science Advisor, Dr. John H. Gibbons, at the Wernher von Braun Lecture, National Air and Space Museum, Washington, D.C., March 22, 1995.

[5]The Motorola Iridium system alone will deploy over 80 satellites in less than two years. The Iridium spacecraft, based on the Lockheed Martin LM-700 bus, is an example of a commercial small spacecraft that can be built in less than 22 days. An estimated 200 commercial small spacecraft will be in orbit by the year 2000.

- What role does advanced technology play in the process of building small spacecraft?

- How should government evaluate civil small spacecraft programs to ensure that objectives are met cost effectively?

2. To identify issues related to NASA's increased reliance on small spacecraft.

3. To provide policy and program-level recommendations based on the research conducted during the study.

Examining small spacecraft programs required a multidisciplinary approach and the study of many dimensions of NASA's programs. RAND's methodology was to rely heavily on close interactions with the NASA offices responsible for conducting space research and the many supporting offices that develop technology and assist in the process of building and testing spacecraft. There were many visits to contractor facilities to review development practices. To help gauge NASA's currently methods, a set of representative small spacecraft missions was selected. Technical data were secured directly from the program office responsible for each of the missions studied. NASA Headquarters' Office of the Chief Financial Officer supplied the cost data. Phone surveys helped to answer specific technical and cost questions. At the midpoint of the study, a workshop was conducted at the RAND offices in Washington, to which engineers and managers from NASA, DoD, industry, and academia were invited. **Because the study's scope was extensive, a large portion of the analysis and many of the conclusions and recommendations are relevant to NASA's programs in the broadest sense.**

The study identifies major issues regarding the cost and performance of small spacecraft missions, reviews the status of technology planning related to these missions, analyzes spacecraft risk and reliability, and suggests strategies for measuring mission performance. Future development trends are also reviewed. Assessing the factors that affect small civil spacecraft may provide useful insights into similar programs being developed in the military space sector.

APPROACH AND METHODOLOGY

To provide a basis for analysis, the study set out to acquire detailed cost and technical data for a variety of small spacecraft. The missions included in this set were

- Discovery: NEAR and Mars Pathfinder

- Explorer: SMEX SWAS and TRACE, and MIDEX MAP

- New Millennium: DS1 and EO1

- SSTI: Lewis and Clark

- Surveyor: MGS, Mars 98 (Lander and Orbiter)

- Clementine.

These missions represent a cross section of NASA's scientific disciplines and a diverse set of design approaches. Some of these spacecraft have already been built and flown; others are still in development. This allowed evolving methods of building and managing programs to be evaluated. Since the mission set included spacecraft built by large and small companies, as well as by NASA's Goddard Space Flight Center (GSFC) and Jet Propulsion Laboratory (JPL), organizational factors could be studied. With the exception of Clementine, the mission set contained only NASA spacecraft. Clementine, a DoD mission designed to test military technologies, was included because of its strong scientific content and objective to test low-cost design and production techniques. The SAMPEX mission, an early example of NASA's return to small spacecraft, was studied but not included in the cost analysis. This was due to the heavy investment that NASA made in SAMPEX as a precursor to follow-on small spacecraft, a factor that would have skewed the cost analysis.

The average dry mass of spacecraft in this mission set was slightly less than 500 kg. To complement the cost data set, a review was conducted of approaches each spacecraft developer took, along with an analysis of the requirements established for each mission.

Although the study focused on civil space missions, several unclassified military programs were reviewed for comparative purposes and to illuminate key aspects of the research.[6] Some larger NASA missions were also reviewed to gain an understanding of the similarities and differences of smaller programs.

The study plan called for an interim workshop. This workshop, entitled "Trends in Development of Small Satellites," was conducted on August 23 and 24, 1996, at RAND's Washington Office. Participants came from OSTP, OMB, NASA, the Air Force, industry, and academia to discuss some of the most recent developments in small spacecraft programs openly. Throughout this report, workshop results are introduced whenever a consensus of the participants was reached. Since the workshop was conducted without attribution, organizational perspectives are not provided.

---

[6]These missions included the Miniature Sensor Technology Integration (MSTI) series, the Space Test Experiment Program (STEP) series, the Radiation Experiment (REX), and the Radar Calibration (RADCAL) mission.

ORGANIZATION OF THE REPORT

This report is structured to answer each of the four study questions in turn. It contains five technical chapters and a final chapter containing conclusions and recommendations.

Chapter Two provides an overview of the cost of small spacecraft missions and their ability to meet challenging scientific requirements. A summary of the cost and technical data gathered and prepared during the course of this study is presented in Appendix A.

Chapter Three investigates the many efforts that have been devoted to building less expensive, more capable spacecraft.

Chapter Four takes a detailed look at the role of advanced technology, as well as NASA efforts to speed the process of maturing and integrating new systems and components.

Process improvement and risk were found to be important aspects of improving performance and reducing the cost of space missions. Chapter Five focuses on process improvement and examines the important role it might play in reducing mission risk. Chapter Five also briefly describes four key technical trends that are especially relevant in terms of the next generation of small spacecraft. A detailed discussion of each of these trends can be found in Appendixes B through E.

Chapter Six addresses the complex issue of measuring mission performance. Finally, study conclusions and recommendations are provided in Chapter Seven. Spacecraft programs are making extensive use of the Internet for information dissemination and coordination. Appendix F provides a World Wide Web (WWW) listing of the sites providing useful information related to small spacecraft.

Chapter Two

SMALL SPACECRAFT IN THE CIVIL SPACE PROGRAM

Small spacecraft have become critically important elements of the civil, military, and commercial space programs. In the space and earth sciences, small spacecraft are performing vital roles in every discipline, as shown in Table 2.1. The importance of small spacecraft warrants a careful examination of their development, suitability, and future capability in terms of meeting our national science objectives in space.

This chapter reviews the role that small spacecraft play in the civil space program and their associated cost and risk. The first section explains the reemergence of small spacecraft within NASA, while the second evaluates the investment the agency has made in this class of spacecraft. The third section reviews the tasks small spacecraft are performing in NASA's science portfolio. The final section estimates the cost effectiveness of these programs. The discussion in this chapter will frame the following chapters of this report.

FACTORS CONTRIBUTING TO THE EMERGENCE OF SMALL SPACECRAFT

The number of small spacecraft in both the military and civilian space programs is growing. The portfolio has changed primarily because the U.S. government has decided to reduce spending in space. Budget reductions translate into a movement away from large, expensive programs and a simultaneous demand for better cost performance. Reducing total mission cost (TMC) is a prevalent theme today within DoD and NASA. Budget pressure is not, however, a complete explanation for the increasing number of small spacecraft. An appreciation of the other factors, such as mission responsiveness and risk mitigation, is needed to place cost reduction in the proper context.

Since the late 80s, NASA has experienced an evolutionary movement back to smaller missions. The Explorer program, NASA's oldest spacecraft series, evolved to Delta-Class missions and is now returning to smaller spacecraft,

Table 2.1

Small Spacecraft Missions

Agency	Office	Program	Representative Missions
NASA	Earth Sciences	New Millennium	Earth Observer (EO) series
		Earth System Science Pathfinders (ESSP)	Pathfinder Series
	Space Physics/		
	Astrophysics	Explorer Program	
		Mid-Sized Explorers (MIDEX)	MAP, IMAGE
		Small Explorers (SMEX)	SAMPEX, SWAS, FAST, WIRE, TRACE
		Student Explorers(STEDI)	SNOE, TERRIERS
		University Explorers(UNEX)	CATSAT
	Planetary	New Millennium	Deep Space (DS) series
		Discovery Program	Mars Pathfinder, NEAR, Lunar Prospector, Stardust
		Surveyor	Mars Global Surveyor, Mars Surveyor '98
DoD	BMDO/USAF	Miniature Sensor Technology Integration	MSTI 1-3
	USAF	Space Test Program	STEP, REX
	BMDO/U.S. Navy	Deep Space Experiment	Clementine

such as the Small Explorers (SMEX).[1] In fact, the SAMPEX mission, launched in 1992, marked the beginning of the turn away from large, expensive spacecraft. A separate evolutionary path, containing the Discovery, New Millennium, and Earth Science Pathfinder series and a set of smaller planetary Surveyors, has emerged recently. From this perspective, NASA's mission portfolio will be dominated, in terms of number of missions, by small spacecraft. This natural evolution of small spacecraft missions will be examined in greater detail in the next section.

Small spacecraft have always been a vital ingredient of the military space program. The primary role of military small spacecraft has been to serve as technology test beds to validate future operational capabilities. In a secondary role, small spacecraft have been used to support basic science objectives. At NASA, the roles of small spacecraft are reversed: The principal application is scientific with a secondary role to test advanced technology, for both the spacecraft bus and the instrument.

It is important to note that NASA has been guided to build "smaller more capable spacecraft to improve the performance and lower the cost of future space missions." (The White House, National Space Policy, 1996, Section 3[d].) This

---

[1]The Delta launch vehicle series has been a popular choice for science spacecraft, commercial communication satellites, and military navigation satellites. NASA statistics show that Deltas have placed more spacecraft in orbit than any other U.S. launch system.

guideline assumes that smaller spacecraft offer clear advantages in terms of technical capabilities and cost-effectiveness, when compared to alternatives. Subsequent sections of this chapter will provide data that do not support this assumption.

The Imperative to Reduce Total Mission Cost

Budget pressure has led to an imperative to cut TMC, which has been driven down into programs to the extent that a new paradigm has emerged within NASA: that "it is enough to return acceptable science; it's not desirable to maximize science by spending more." (NASA Astrophysics Subcommittee, 1994, p. 12) Science is no longer the overwhelming driver in mission design. Managers have always had to consider science, cost, and technical requirements when designing a mission, but these parameters now carry more or less equal weight.

New NASA Announcements of Opportunity (AOs) are being evaluated on the basis of scientific and technical merit *and* the total proposed cost. Each proposal must establish a "Baseline Mission," the mission that returns the full set of science objectives, along with a "Minimum Science Mission," beneath which the investment would no longer be justified. "Descope options" must be prepared that outline a staged reduction to the minimum science mission.[2] The descope feature is a reflection of "design-to-cost" engineering. A prescribed descope path allows the spacecraft developer to respond quickly to problems in design and fabrication.

The need to reduce the TMC of space science missions is accelerating the shift to small spacecraft. Yet building smaller, less expensive spacecraft is only one way of responding to tightening budgets: Why not simply stretch programs? Are there other pressures on NASA's science programs?

Responding to the Needs of Science

A vigorous debate over the relative merits of large versus small spacecraft has existed within NASA and the space-science community since the inception of the space program. (Newell, 1980, p. 124.) By the mid-80s, support was building for a greater number of smaller missions. Well before the current climate of budget constraint, the need for change within the space-science program was documented in several landmark advisory reports. In a 1986 report entitled *The Crisis in Space and Earth Science*, the Space and Earth Science Advisory Com-

---

[2]Examples are the Mid-Sized Explorer (MIDEX) Program AO, December 1994, and, more recently, the Earth System Science Pathfinder (ESSP) AO, July 1996.

mittee (SESAC) determined that "NASA should reexamine its approach toward implementation of flight projects with the intent of reducing overall mission cost." The report further noted that "rapid, elegant response is imperative," suggesting that "science is not best served by exclusive emphasis on major missions." (SESAC, 1986, p. 28.)

During this period, the long lead-times and cost and schedule growth of many major missions, such as HST and CGRO, had choked new starts. Many senior officials, looking downstream at the changing political climate, felt "that NASA was in danger of pricing itself right out of business." (Hamaker, 1991, p. 7.)

That program-development timelines had become unresponsive to the needs of the science community was further underscored in a 1993 National Research Council (NRC) report:

> Efficient conduct of science and applications missions cannot be based solely upon intermittent, very large missions that require 10–20 years to complete. Mission time constants must be commensurate with the time constants of scientific understanding, competitive technological advances, and inherent changes in the systems under study. . . . NASA's new initiative for smaller, less expensive, and more frequent missions is not simply a response to budget pressures; it is a scientific and technical imperative. (NRC, 1993, p. 3.)

It was clear that, in this environment, smaller, fast-turnaround missions were needed. Further, these missions had to step beyond the traditional pathfinder role they were relegated to during the 70s and 80s. Synergism with larger missions emerged as the new NASA strategy.

Responsiveness to science was a role well suited to smaller scientific platforms. The responsiveness of small spacecraft to mission requirements depended, however, upon the space-science discipline being examined. Fortunately, many aspects of space physics and astronomy, such as sun-earth dynamics and background-radiation experiments, could make ready use of small spacecraft because of instrument availability.

Under heavy pressure from the space-science community to refocus the program, NASA's Office of Space Science and Applications (OSSA)—the predecessor of the current Office of Space Science (OSS)—created a detailed Strategic Plan in 1988. This plan contained a long-range vision to balance the space-science program, including clear criteria for prioritizing, funding, and controlling the cost of missions. Beginning with space physics, the plan outlined a strategy to reintroduce small spacecraft into the NASA portfolio, noting that they could be

> built and launched within three years, yet they are sufficiently capable to accomplish first-class scientific objectives in astronomy, space physics, and upper atmospheric physics. (OSSA, 1988, p. 16.)

The decision to utilize small platforms *as mainline spacecraft* in the mission portfolio was made after careful review of technological capabilities and DoD experience with the use of low-cost, low-mass platforms.[3] An additional benefit of smaller missions was that they "stimulate the research community, particularly at universities, with exciting new opportunities, which will attract new scientists and engineers to space science." (OSSA, 1988, p. 16.)

In presenting a long-range vision for space and earth science, the 1988 OSSA Strategic Plan was a pathfinding document, but it did not foresee impending budget reductions. It did, however, establish a long-range planning culture within the science community. In so doing, the plan set the stage for a significant small spacecraft contribution, helping to prepare the community for the current budget environment.

Today there is a strong synergism between small and large spacecraft. The WIRE spacecraft's instrument, for example, uses mid-infrared (IR) sensors being developed as part of the SIRTF Advanced Technology Development (ATD) program. (NASA Astrophysics Subcommittee, 1994.) Small spacecraft have also evolved to exploit opportunities identified by larger missions. The evolutionary development of small spacecraft can be seen in the recently announced MIDEX Microwave Anisotropy Probe (MAP). The residual microwave background, first discovered by Wilson and Penzias at the Bell Laboratories in 1965, was explored extensively using the Cosmic Background Explorer (COBE) (2,140 kg). One of the principal discoveries of that mission was anisotropy in what was previously thought to be a uniform microwave profile. The MAP (553 kg) mission will complete a survey of the cosmos with a sensitivity two orders of magnitude greater than COBE.

In the planetary sciences small spacecraft are expected to revitalize research programs that had become stalled by infrequent flight opportunities. (Lawler, 1997, p. 1596.) A recent COMPLEX report concluded that

> we have finished the preliminary reconnaissance of the major bodies in the solar system and have entered an era of intensive study of the physical phenomena that shape our planetary neighbors . . . [T]he initiation of a series of small missions presents the planetary science community with the opportunity to expand the scope of its activities and to develop the potential and inventiveness of its members in ways not possible within the confines of large, traditional programs. (NRC SSB, 1995.)

---

[3]The 1988 OSSA Strategic Plan did not address small planetary spacecraft, mainly because the needed miniaturized instrumentation was unavailable. Technology has matured rapidly in this area, enabling small missions to the planets. Small, lightweight power sources, such as scaled-down nuclear and ultraefficient solar power systems, are not yet available, a factor currently limiting small spacecraft much beyond the orbit of Mars.

The initial set of small spacecraft for planetary missions has had fairly modest measurement objectives. Examples of such missions are DoD's Clementine technology demonstrator spacecraft (a lunar mission with a planned flyby of the Geographos asteroid) and NASA's Discovery Near Earth Asteroid Rendezvous (NEAR), Mars Pathfinder (MP), Stardust, and Lunar Prospector (LP) spacecraft and the Surveyor Mars '98 Orbiter and Lander.

Taking Advantage of New Technology

Advanced technologies do not automatically lead to small spacecraft. Conversely, spacecraft can be made smaller without advanced technology. However, a spacecraft designer asked to deliver high performance in a smaller, less-expensive package is driven to incorporate advanced systems. Delivering more performance from small spacecraft is foremost in the minds of mission managers. Despite budget pressures, NASA must produce "world-class" results to maintain a viable program endorsed by the science community.

Budget reductions have led to a substantive reduction in the content of NASA's science programs. Advanced technology, however, provides a means to recapture science content through increased performance, while also offering the potential for lower costs. The current generation of NASA missions approaches the state of the art more closely than earlier ones. Here, small spacecraft offer a distinct advantage in that their shorter development timelines allow incorporation of the latest systems and components. Features of advanced technology programs and the importance of adequately planning them will be addressed at length in Chapter Four.

NASA'S INVESTMENT IN SMALL SPACECRAFT

Missions much larger than the class examined in this study are unlikely to reappear in the near future. Figure 2.1 presents a projected budget for NASA through the year 2002 (not reflecting full-cost accounting), showing an essentially flat allocation for space- and earth-science programs. Earlier budget projections called for deeper reductions in the agency's science budgets, creating urgency in the shift to smaller programs. In 1995, for example, OSS faced a 25-percent decrease (from $2 billion in FY95 to $1.54 billion in FY00) over five years (Huntress, 1995a). The flat budget profile for science programs is, of course, an effective decrease of approximately 15 percent in purchasing power by 2002. Discretionary programs must compete among themselves within the flat Federal budget. NASA's hard commitments—to the International Space Station Program, the Space Shuttle, and to the technology for eventual replacement of the Orbiter fleet—will likely prevent much growth in funding for science missions.

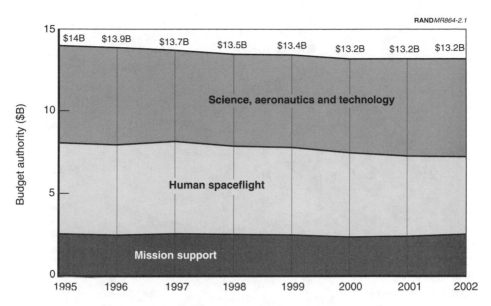

SOURCE: NASA Headquarters (1997), pp. 511–513.

Figure 2.1—NASA Budget Projection Through 2000 (real-year dollars)

Satellite-based research and application programs represent a significant percentage of civil space funding. Figure 2.2 examines NASA spending in FY96 for (1) all spacecraft research programs and (2) for small spacecraft programs within the scope of this study.[4] Cost breakdowns are provided in five areas: hardware (flight systems development), launch systems, operations, R&A, and personnel. Approximately $4 billion are spent on spacecraft research programs within NASA, with $1 billion devoted to small spacecraft missions.

Small spacecraft, therefore, represent a significant Federal investment, a percentage that will grow as larger spacecraft, still under development, are completed. As shown in Figure 2.3, the mass of research spacecraft has dropped dramatically in recent years. The large missions remaining on the chart are, for the most part, legacy programs that were initiated prior to sharp budget reductions.

_____

[4]This is a crosscut of the NASA budget prepared using RAND's Research and Development in the United States (RaDiUS) database, which contains detailed information on the Federal budget. Spacecraft costs are a summation of development costs for the flight segments of the respective cost cuts. Launch costs include vehicle procurement and spacecraft-to-vehicle integration costs. Operational costs represent an aggregate of facility costs, individual mission operations and data analysis (MO&DA) costs, construction of facilities (CoF), and ground-segment line items. Research and analysis (R&A) is a simple accumulation of these identified line items. Personnel costs are estimated as a fixed percentage of overall research and program management (R&PM) accounts.

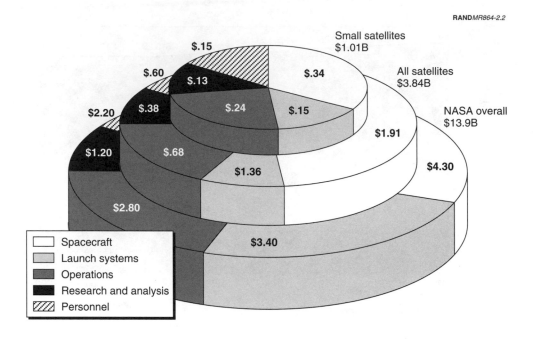

Figure 2.2—Satellite Budgets at NASA (FY96)

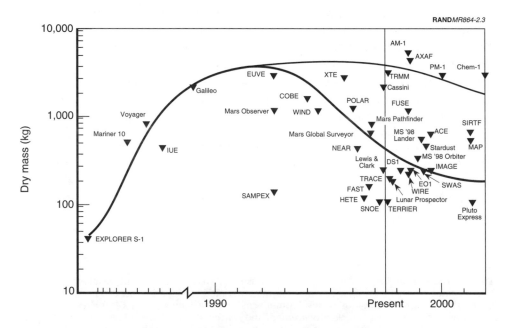

Figure 2.3—Mass Trends in NASA Satellite Missions

THE ROLE OF SMALL SCIENTIFIC SPACECRAFT

The amount of science lost due to budget reductions and a shift to smaller spacecraft is difficult to estimate. However, a careful examination of advisory committee reports, discussions with mission scientists and engineers, and other relevant factors yield several important observations.

The growing complexity and cost of large spacecraft programs, along with fear of budget reduction, led NASA to a substantial restructuring of some programs and the cancellation or postponement of many others.[5] The gradual shift to smaller, leaner programs accelerated when budgets began to shrink in FYs 94 and 95. As the portfolio changed to include more small spacecraft, the scientific content of the portfolio also changed. That the content of NASA's space- and earth-science portfolio has diminished is reflected in Figure 2.4, taken from OSS's Strategic Plan. Advanced technology is seen as the mechanism for allowing smaller, less-expensive spacecraft to reestablish program content. Funding for advanced technology, however, must be found within an essentially flat budget.

The small spacecraft that have been built up to this point have clearly demonstrated an ability to deliver first-class science, but they cannot yet, and were not expected to, deliver the same science for less money. Additionally, there is some concern that shortfalls in the out-year budgets for NASA's science program could produce missions that cannot return useful science.[6] Of particular concern is the phasing of missions. Many science missions contain highly interactive components. The utility of observational spacecraft, for example, is heightened if particular instruments are on orbit at the same time for coordinated viewing. (NRC, 1996, p. 3.)

RAND workshop participants also felt that small spacecraft show clear promise for the future but do not replace the need for larger missions; the Small Explorer Wide-Field Infrared Explorer (SMEX-WIRE), is a case in point. Advanced spacecraft and instrument technologies are allowing small spacecraft to deliver impressive results. WIRE will observe sources 500 to 2,000 times fainter than could the Infrared Astronomical Satellite (IRAS), which was launched in 1983 weighing 1,100 kg.[7] Although demonstrating extraordinary performance for a

---

[5]Examples are the Advanced X-Ray Astrophysics Facility (AXAF) and the Far Ultraviolet Spectroscopic Explorer (FUSE), which were heavily restructured; the Comet Rendezvous and Flyby (CRAF), which was canceled; and the Solar Probe (SP) and the Space Infrared Telescope Facility (SIRTF), which were deferred.

[6]Several NASA advisory committees have expressed concern that returns from the planned program may prove inadequate to support the space-science community. Examples are the Minutes of the Space Science Advisory Committee, May 15, 1996, and a recent NRC review. (NRC SSB, 1996a.)

[7]Minutes of the NASA Astrophysics Subcommittee (1994).

RAND*MR864-2.4*

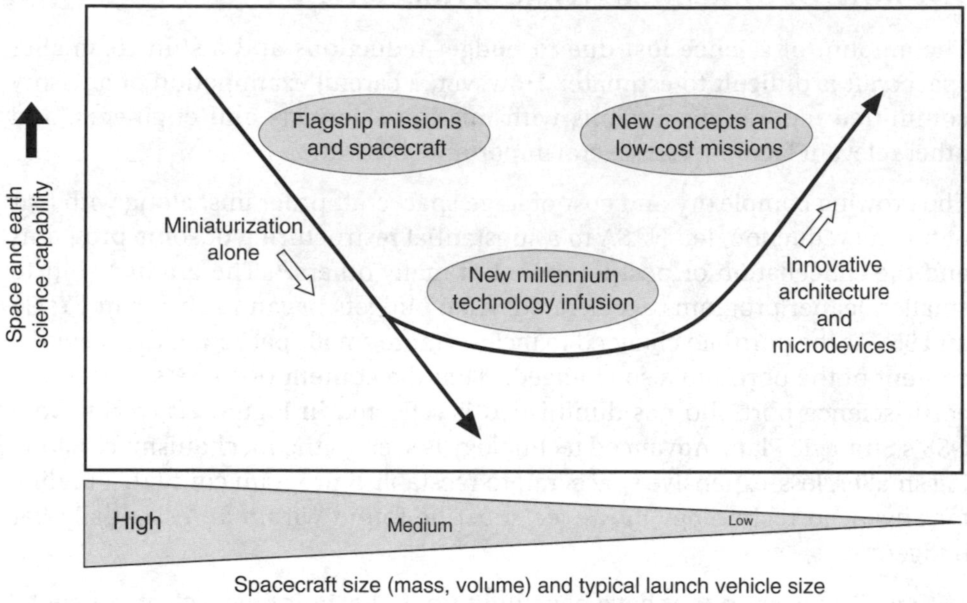

SOURCE: OSS (1995b), p. 25.

Figure 2.4—Road Map for Microspacecraft Development

spacecraft with a development cost of approximately $50 million, WIRE will observe only a limited portion of the IR spectrum planned to be mapped by the larger SIRTF mission.

There are additional indications that small spacecraft are not yet the ideal solution in all applications; plans for Mars exploration are illustrative. NASA's plan is aggressive, calling for spacecraft to be launched at windows occurring in 1998, 2001, and 2003. But available funding may not be sufficient to send spacecraft with adequate instrumentation. The NRC's Committee on Planetary and Lunar Exploration (COMPLEX) concluded that the planned series of Mars missions was limited in its ability to perform the types of investigations needed by the scientific community. (See NRC SSB, 1995; NRC SSB, 1996a.) When asked how best to optimize Mars research, scientists at a 1996 NRC Workshop on Reducing Mission Cost, echoing the COMPLEX conclusion, questioned the assumption that a small orbiter-and-lander mission for the 2001 opportunity was preferable to applying funds to a larger spacecraft for the 2003 launch window. (NRC, 1996, p. 15.)

Most small spacecraft builders agree that small spacecraft cannot yet meet all of the prime space- and earth-science objectives. Rather than replace their larger counterparts, small spacecraft currently exploit opportunities that have been identified by previous missions, perform focused investigations, and serve in a precursor role. RAND workshop participants urged the development of appropriately sized spacecraft that represent the best balance of scientific return and available funding.

THE COST EFFECTIVENESS OF SMALL SPACECRAFT

A considerable degree of mystique surrounds small spacecraft. Some of this is due to the fact that many small spacecraft are proving that they can return a great deal of scientific data. On the other hand, the cost of small spacecraft has sometimes been oversold. Since small spacecraft can be built more quickly, they are intrinsically less expensive. They therefore hold natural appeal in a budget environment that can no longer afford larger ones.

Beyond the naturally lower cost of small spacecraft, it is important to ascertain whether they offer cost efficiencies. Figure 2.5 shows spacecraft development cost relative to dry mass. The calculation includes the cost to manage, design, develop, and test *the spacecraft and instrument* and excludes any launch, ground-support equipment, and operational costs. As shown in Figure 2.5, there is a wide variation in the relative cost of smaller spacecraft, and they can cost appreciably more per kilogram than larger ones. The linear regression line shown in Figure 2.5 was drawn using missions where the spacecraft cost was

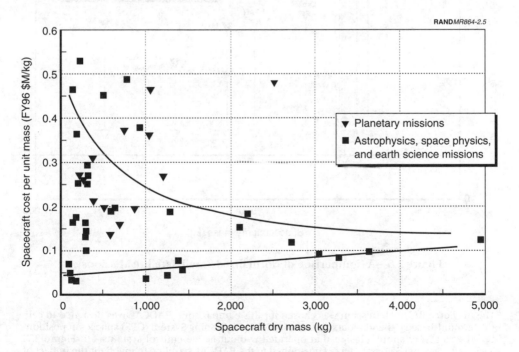

Figure 2.5—The Relative Costs of Small Spacecraft

less that $100,000 per kilogram. The lower-mass end of this trendline represents such missions as NASA's STEDI and UNEX, which are built with existing technology using straightforward design and fabrication techniques. NASA's mainline science spacecraft, including the missions reviewed in this study, are, however, best represented by the exponential trendline shown. Planetary missions were plotted separately in Figure 2.5. Although there is appreciable scatter in the data, smaller planetary spacecraft do appear to cost less than larger missions because, in some part, of their ability to use simpler power and propulsion systems than their larger counterparts.

Increasing complexity was the principal factor driving the cost of larger missions to the point where they were untenable. (Hamaker, 1991, p. 7.) The increased variation in cost-per-kilogram at lower masses is also due in large part to differences in spacecraft complexity. The effect of complexity is illustrated in Figure 2.6. In this chart, the relative spacecraft development costs of the missions evaluated in this study were normalized to account for complexity. This was accomplished by comparing them to a simple, low-cost mission.[8]

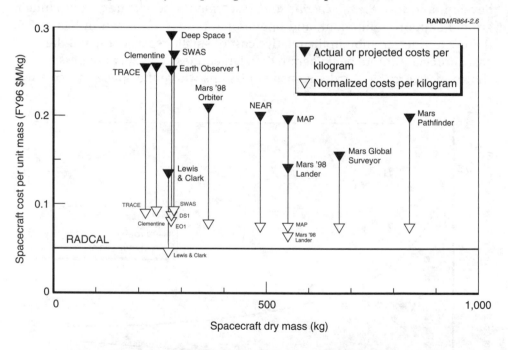

Figure 2.6—A Comparison of Normalized Small Satellite Missions

---

[8]The Air Force RADCAL mission was chosen for this comparison. RADCAL was designed to calibrate ground-based C-Band radars and to test Global Positioning System (GPS) spacecraft positioning. It was a very simple design that operated without active control systems. In Figure 2.7, spacecraft costs-per-kilogram were constrained to the RADCAL baseline to highlight the impact of complexity on cost. Details on the calculation of complexity are presented in Appendix A.

Cost models contain extensive parametric analyses to establish the sensitivity of spacecraft cost to such variables as pointing accuracy and power generation.[9] The complexity factor used in Figure 2.7 is a first-order calculation that evenly weights the many variables that determine mission cost. The accuracy of Figure 2.7 is sufficient to demonstrate the importance of considering complexity when comparing the relative costs of missions. It also illustrates that reduction of complexity is an important strategy for controlling costs. Many small spacecraft are relatively expensive because they retain the complexity required to attain demanding science objectives.

But there are other factors that determine the cost of a spacecraft. The *class* of a spacecraft has a strong influence on cost.[10] The size of the team and the approach used to design and develop the spacecraft also have a strong influence on cost.

The risk of failure also affects cost. To keep costs down, many small spacecraft missions have pursued high-risk development approaches:

> There is intensive pressure on the new faster, better, cheaper programs to provide extraordinary levels of science and complexity while staying within severely constrained schedules and budgets. The mission risks are increasing, and upper management must be made fully aware of the risks involved. (Brown et al., 1996, p. 3.)

Within available funding, every attempt is made ensure mission reliability, but the increased potential for partial or complete failure must be recognized. If a spacecraft is lost, another is usually constructed and launched. Even a partial failure translates into lost opportunity costs. To the extent that a spacecraft mission incorporates more risk, it has a higher *cost of failure*.[11] The subject of risk is covered in greater detail in Chapter Five.

---

[9]Most aerospace cost models contain few empirical data on the development cost of small spacecraft. The reader is directed to two models that deal explicitly with these systems: the Small Spacecraft Subsystem-Level Cost Model, developed by the Aerospace Corporation, and SMALLSAT, a Phase A design tool relying on similar cost data, developed by Princeton Synergetics, Inc. Also, NASA GSFC's Resource Analysis Office (RAO) and JPL's Engineering and Science Directorate maintain detailed data on past mission costs. Finally, the NASA/Air Force Cost Model (NAFCOM) contains data on many small missions.

[10]The "class" of a spacecraft refers to the standards and controls used in its construction. "Class A" refers mainly to human-rated spacecraft. At the other end of the spectrum, a "Class D" spacecraft can be built using commercial-grade components with relaxed inspection and test standards. The majority of small science spacecraft are built to the equivalent of a "Class C" standard. Such spacecraft as the X-Ray Timing Explorer (XTE) have been built to these standards without suffering loss of reliability or performance. The reader is referred to NASA Management Instruction (NMI) 8010.1A Appendix A (now expired), for a detailed definition of spacecraft classifications.

[11]A more detailed discussion of cost of failure can be found in Hecht (1992), pp. 700–704.

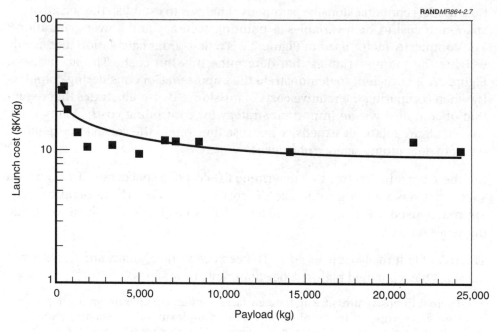

SOURCE: S. Isakowitz (1994).

Figure 2.7—Trends in the Specific Cost of Low-Earth Orbit Launchers

Small missions also pay proportionately higher prices for launch services. Although smaller, lighter-weight spacecraft can use less expensive launch vehicles, this cost advantage is somewhat offset by the higher relative cost of small launchers. As shown in Figure 2.7, there is a significant economy of scale in the launch market. Market pressures offer few solutions, since commercial providers can only be expected to set prices below competitor levels. Prices for launch service could be expected to drop significantly only if there were adequate competition in the market. There are, however, few funded programs that could hope to change the launch economy in the immediate future. The impact of launch cost reduction on TMC is also modest for small missions. For example, for the missions reviewed in this study, launch costs accounted for 20 percent of TMC; a 25-percent cut in launch cost would reduce TMC by only 5 percent.

It can be seen that the types of small spacecraft needed to fulfill today's science requirements are complex devices. Being smaller, they require less time and effort to build and are, therefore, cheaper. On a per kilogram basis, however, they cost more to build and launch. To the extent they incorporate higher levels of risk, they also carry a greater cost of failure.

SUMMARY

The movement to smaller spacecraft began in response to the science community's call for more frequent, less costly missions. The shift to smaller missions accelerated because of the budget pressures of the early 90s, a constraint that will likely remain for the foreseeable future.

Small spacecraft have become a critical component of NASA's space- and earth-science programs. They provide an important response (a) to the demands of the science community for faster and more frequent flight opportunities, (b) to national budget priorities, and (c) to the desire to distribute risk. Though they are capable of impressive levels of performance, the current generation of small spacecraft cannot, and was not intended to, produce a science program equal in content to past programs. They are, however, ideally suited to meeting niche mission requirements, often providing insights into a particular phenomenon or paving the way for broader-scale investigations.

It is certainly true that small spacecraft can be built faster and cheaper, but this is hardly a surprising result. More relevant is the fact that most small spacecraft do not offer a clear cost advantage over larger systems. To meet demanding mission requirements, small spacecraft remain complex, and complexity costs money. Economies of scale in the launch market also mean that it is relatively more expensive to lift smaller spacecraft to orbit. To the extent that they represent greater technical risk, the overall cost of the space and earth science portfolios could be higher. With these considerations in mind, small spacecraft should be carefully integrated into NASA's science programs. Their cost and utility should be balanced against scientific requirements to create a program that optimizes returns from available funding.

The National Space Policy should be reviewed and revised to ensure that NASA is guided to plan and implement appropriately sized missions. The focus on the policy should be on achieving scientific objectives at the best balance of cost and risk.

Finally, it is important to consider that the small spacecraft economy could change substantially if performance continues to increase while development costs drop. Both conditions might be met through the adoption of advanced technology and improved design and development processes, especially if such initiatives reduce spacecraft complexity. Technology efforts and process improvement will be discussed in future chapters, following a discussion of what already has been done to reduce mission costs.

Chapter Three

MEETING NATIONAL OBJECTIVES WITH FEWER RESOURCES

RAND was asked to review and comment on the many strategies NASA, industry, and academia use to reduce cost and increase performance. The first section of this chapter presents a review of improvements made in the areas of management, procurement, design, operations, and standardization. The next section examines the effect of a maturing commercial sector on cost and performance. The third section examines the effect of process improvement initiatives and quality practices on improving management and production efficiency. The fourth section presents a discussion of some of the important hidden costs associated with the shift to smaller programs. The chapter concludes with an estimation of the savings that have accrued from these various initiatives.

COST-REDUCTION STRATEGIES

Recently, extensive research has been focused on the factors driving the cost of building and operating spacecraft. Noteworthy has been the willingness, on the part of spacecraft builders, to share information from lessons learned on past missions. Workshops and conferences dedicated to spacecraft construction are in abundance. The WWW is also expanding the communication pathways open to designers and scientists alike.[1] Seen in this light, the pressure to reduce mission costs has spurred innovation and the rapid expansion of formal and ad hoc communication networks.

In regard to small spacecraft, DoD and a host of foreign, university, and amateur builders have built an extraordinary variety of small spacecraft over more than three decades. As a result, a great deal has been published on methods of reducing the cost of these missions.[2] It is important to remember that much, if

---

[1]See Appendix F for a list of small satellite Internet web addresses. Many of these sites use the Internet to coordinate the development team.

[2]A treatise on these methods can be found in Wertz and Larson (1996). Also useful are the *Proceedings of the Annual AIAA/Utah State University Conference on Small Satellites*, and the 1996 Goddard Workshop on Small Satellites. Information on these events is available at the Utah State University

not most, of what has been learned during cost-reduction exercises is applicable to spacecraft development programs of any size. Future management and engineering processes will benefit from the techniques that are fundamentally changing the spacecraft development culture.

Streamlined Procurement Practices

Management by performance measurement rather than specification is the best way to characterize the new approach to Federal procurement.[3] NASA has evolved from a dependence on cost-plus-fixed-fee (CPFF) to award-fee contracts and later to a predominance of incentive-fee contracts. Lately, the trend has been toward fixed-price contracts.

There is also a growing reliance on "contractor best practices" instead of mandated inspection and quality standards. These decisions acknowledge the prior investment companies have made to certify personnel and facilities to meet historical spacecraft standards. As documentation requirements have been reduced, so has a dependency on paper exchange. A greater percentage of documents are exchanged electronically, taking advantage of the Internet and a greater degree of document format standardization.

Many of the acquisition strategies being used to great advantage within small spacecraft can be traced to streamlining of the AO process within NASA. Any procurement vehicle can be used, but recent AOs provide cost caps, which invite development of spacecraft under fixed-price or performance-based contracts.

Fixed-price and performance-based contracts are not new in the context of small spacecraft, having been used since 1965 in the Air Force's Space Test Program (STP). The Air Force views STP as a testing ground for new spacecraft technology and for experimenting with new forms of government-contractor relationships. For many years, STP has operated under streamlined procurement rules. For example, STP spacecraft are typically built using 22 contract document requirement lists (CDRLs) instead of the hundreds required on traditional Air Force and NASA programs.

The Air Force has also experimented with "packaged buys," in which the contractor provides not only the spacecraft and instrument but also the launch

---

and GSFC WWW sites listed in Appendix F. Data related to current small satellite programs can be found in Gipson and Buenneke (1992). Finally, the development of very low-cost small spacecraft is covered in Fleeter (1995).

[3]Small spacecraft programs have benefited from NASA, DoD and commercial procurement streamlining practices and paperwork reduction initiatives. NASA procurement reform efforts were reviewed in a 1992 House Committee on Science, Space and Technology Hearing Report, HR-108. See also Bowers and Dertouzos (1994).

vehicle and delivery-to-orbit operations. Fees and incentives are determined by demonstration of on-orbit operation. This is essentially the path that NASA's Discovery and SSTI programs have taken.

STP also explored the application of standard small buses, purchased under "variation in quantity" contracts. An example is the Space Test Experiment (STE) Program series of small spacecraft built by TRW, Inc. (see Figure 3.1). Each spacecraft bus unit in the purchase was considered a fixed-price delivery. The STEP experience illustrates the need to limit changes within the spacecraft design and integration process strictly. Increases of 18 to 20 percent occurred above the "core price" of the basic STEP spacecraft, mainly because of changes

SOURCE: Department of the Air Force (SMC/PAS)

Figure 3.1—The STEP-04 Spacecraft

in the configuration of the payload, which required subsequent modification of the bus.[4] These variations, however, still allowed the STEP series to demonstrate cost efficiencies over traditional methods of procuring spacecraft.

Additional attention is being focused on use of standard commercial spacecraft buses through the NextGen satellite initiative (being jointly pursued by the National Reconnaissance Office, NASA, the Air Force, and the Navy) and by efforts at NASA Goddard to establish "indefinite design, indefinite quantity" (IDIQ) contracts with existing bus suppliers.[5] These efforts seek to exploit the availability of inexpensive commercial systems, purchasing existing designs at marginal costs. NASA is also evaluating the possibility of expanding the planned IDIQ procurement under Federal Acquisition Regulation (FAR) Part 12 rules.[6]

Low-Overhead Management Techniques

Cost and/or schedule growth has been a historical part of spacecraft development programs. Table 3.1 depicts this trend in some past programs. This growth can be largely attributed to the risk-avoidance practices of the past, to premature commitment to designs that had not been adequately fleshed out, and to hierarchical organizational structures that provided insufficient management visibility into the development process. Management reform, on both the development side and the science requirements side, has been an important step in reducing spacecraft TMC and preventing cost and schedule growth.

New design and risk-management processes have greatly reduced the likelihood of overruns. Increased reliance on fixed-price contracts (FAR Part 12 mandates the use of fixed-price contracts) and performance-based procurements is also greatly reducing the risk of cost and schedule slips in small spacecraft programs, as is a NASA commitment to the cost caps outlined in the AOs. Emphasis on cost and schedule accounting is prevalent within small spacecraft programs. Widespread use of commercial-off-the-shelf (COTS) financial management and planning software programs has allowed mission managers to gain insight into progress and make decisions early. Managers are also emphasizing lean development teams, relying on matrix organizational structures to a

---

[4]Site visit, TRW East Coast Operations, February 1997.

[5]"U.S. Agencies Work to Form Next-Generation Satellites" (1997); Brown (1997).

[6]The heart of the 1994 Federal Acquisition and Streamlining Act (FASA) is reflected in FAR Part 12, Acquisition of Commercial Items. FAR Part 12 defines a commercial item as "any item that is of a type customarily used by the general public or by non-governmental entities for other than governmental purposes." The DoD has broadly interpreted FAR Part 12 to include aircraft, such as the C-130J, which is similar to types sold to commercial airlines.

Table 3.1

Schedule and Cost Growth for Select Space Science Spacecraft

Mission	Mass (kg)	Launch date	Delivery (months)	Schedule Growth (percent)[a]	Cost Growth (percent)[a]
AE-C	50	1973	17[b]	47	16
HEAO-A	1,182	1977	43[b]	71	21
AEM-HCMM	84	1978	24[b]	n/a	n/a
Magellan	1,035	1989	72[a]	24	84
Galileo	2,380	1989	n/a	195	157
CGRO	15,700	1991	118[a]	78	205
Mars Observer	1,018	1992	n/a	32	83

SOURCES:
[a]Tyson (1992).
[b]Harmon (1993).

greater extent. Groups that enjoy access to larger corporate resources, for example, will draw upon expert consultants from the parent organization on an as-needed basis. This helps to ensure that senior experience is brought into smaller programs.

Information systems are having a significant influence on the ability of small spacecraft to coordinate activities, not only in the sense of reporting but also in terms of exchanging design information. In some cases, engineers are using information systems to access remote design tools and operate testbeds.

The traditional Phase A–E milestones of spacecraft programs have also been, to a significant degree, abandoned by small spacecraft teams in favor of ongoing in-process reviews. Status meetings are frequent, with teams unwilling to wait until a certain schedule milestone to discuss problems or potential problems. This more fluid approach provides managers with better visibility into the progress of a project, permitting early intervention into emerging problems.

Management of science missions increasingly rests with a principal investigator (PI). In the "PI Mode," an investigator is selected who has full responsibility

> for all aspects of the mission, including instrument and spacecraft definition, development, integration, and test; launch services (if acquired by mission team) or mission launch interfaces (if launch service is NASA provided); ground system; science operations; mission operations; and data processing and distribution.[7]

---

[7]NASA Office of the Mission to Planet Earth (1996), p. 18.

The PI need not be a NASA scientist, and the team selected may or may not consist of NASA elements. It is likely that the majority of NASA space- and earth-science missions will be managed in PI mode. The Explorer program, for example, will exclusively rely on PI-mode management for future missions. (NRC SSB, 1997, p. 18.)

PI-mode management is an important development that allows NASA to maintain leaner internal organizations. Agency spacecraft-development teams can compete to participate in these missions, but there is reason to wonder whether the shift to external sources for the management and implementation of missions will lead to erosion of NASA's core competencies. Further, following the guidelines of the National Space Policy, NASA will develop spacecraft only when the unique technical capabilities of a center are required.[8] The ability to procure components, systems, and entire spacecraft buses from a maturing private sector presumably means that there is less for government engineers and managers to do. Transfer of technical and management functions to the private sector and academia could mean substantial NASA staff reductions. (Musser, 1995.)

NASA scientists and engineers can compete for PI-management roles, and agency spacecraft groups are not precluded from leading, or being part of, a winning mission proposal. NASA's shift to full cost accounting could influence the ability of in-house elements to compete, however.[9] Historically, the agency reported only direct costs (those clearly or physically linked to a specific project) when describing a mission. The cost of civil-servant labor was not historically reported. NASA built three of the missions reviewed in this study in house. Labor alone increased the reported cost of these spacecraft by an average 25 percent. Full-cost procedures add such elements as general and administrative, service, and facility charges.[10] When summed, the costs of NASA-built space-

---

[8]Under Part (4)(c) of the 1996 National Space Policy, NASA is guided to acquire spacecraft from the private sector, unless the NASA Administrator judges that the expertise of a NASA field center is required. Under NASA's 1995 Zero-Base Review (ZBR) guidelines, only GSFC and JPL are permitted to build in-house science spacecraft. These guidelines limit in-house development to "first-of-a-kind technology." It is not clear, however, whether this definition provides sufficient precision when choosing where a spacecraft will be built, since each science spacecraft is sufficiently unique to be considered first of a kind; see W. Huntress (1995b).

[9]Legislative pressure to reform Federal accounting systems is reflected in the 1990 Chief Financial Officers Act, 1993 Government Performance and Results Act (GPRA), and 1996 Federal Financial Management Improvement Act. The 1993 National Performance Review, and NASA's 1995 Zero Base Review and Federal Laboratory Review, each highlighted the need for full-cost data. Under the Integrated Financial Management Plan (IFMP), NASA plans to implement full-cost accounting in stages, with completion scheduled for FY 2000.

[10]Proposed full-cost accounting methods are provided in NASA Office of the Chief Financial Officer (1997b).

and earth-science missions could rise to 40 percent above those historically reported.[11]

Finally, there is some evidence that part of the streamlining that has taken place at NASA involves relearning how to build fast-turnaround missions. For example, during the 70s, the Atmospheric Explorer C (AE-C) and Application Explorer Mission–Heat Capacity Mapping Mission (AEM-HCMM) spacecraft, listed on Table 3.1, were built in 17 and 24 months, respectively, measured from the initial fabrication contracts to the delivery of the spacecraft for launch. This is roughly equivalent to what is being achieved today. It is perhaps worth remembering that the first Thor missile flew only 13 months after contract award (Worden, 1994).

Design and Development Improvements

To reduce mission costs, managers must carefully monitor design activities. High-cost engineering hours must be minimized, and the design phase must be as brief as practical. This often means shifting emphasis from design to integration and test of the spacecraft. The philosophy that "we'll fix it in I&T" was found consistently among small spacecraft builders.

A small spacecraft program manager must balance pressures to control design costs against a need to integrate new, high-performance technologies whenever possible. This demands careful management in other areas. Spacecraft managers seek to obtain early commitment to requirements and interface specifications, which are then closely controlled to prevent creep.

Designs are reused whenever practical to further control costs. For example, approximately 85 percent of the software from the XTE was reused in the Tropical Rainfall Mapping Mission (TRMM). (Ryschkewitsch, 1995.) Such levels of reuse are historically unusual, but small spacecraft are evolving along a path that supports this practice. Spacecraft series, such as the Explorer missions, are pursuing "scalable" designs. This helps to ensure design reuse between large and small spacecraft. Spacecraft built in a series also have the advantage of team stability. Lessons learned from each mission can be captured quickly and reflected in follow-on designs.

Streamlining spacecraft design and development processes too much can lead to the loss of critical elements. Some of the lessons learned from the Wakeshield spacecraft illustrate this. Wakeshield, shown in Figure 3.2, is a 2,000-kg Shuttle-deployed free flyer designed to produce ultra-low vacuums for

---

[11]NASA Office of the Chief Financial Officer (1997a), p. 4.3.3.

SOURCE: JPL.

Figure 3.2—Wakeshield Spacecraft in Flight

space-based materials processing. The spacecraft was built using a strategy of very low production costs, minimal documentation, and reliance on single-string architectures. The initial two flights of Wakeshield were less than successful. The first flight in 1994 was terminated prior to release when communication problems prevented command signals from reliably reaching the spacecraft. A second 1995 flight was partially successful, but electrical interference originating on the spacecraft caused attitude variations deemed unacceptable to the processing objectives of the mission.

Following these initial two missions, it was decided that the basic design of Wakeshield was sound but that elements of the spacecraft needed to be reinforced and rebuilt. Redundancy was added to the communication system; batteries were added to provide more power margin; the thermal system was augmented; and the attitude control system was greatly improved. In performing these augmentations, the Wakeshield team found it difficult to track and resolve problems because documentation was often sparse and not uniformly controlled. To save money, an independent test team had not originally been formed, so design engineers also had to perform this function. This seriously overloaded the engineering team, whose familiarity with the spacecraft sometimes prevented a dispassionate overview of problems. Finally, the lack of a formal system engineering function led to problems late in the development cycle, when they were costly to fix. These issues were overcome, and the third flight of Wakeshield in November of 1996 was successful.[12]

Lowering the Cost of Operations

Advances in technology are allowing great strides to occur in the operation of small spacecraft. These advances are found both on the spacecraft and on the ground. Processors with greater computational power, combined with fault detection and health-monitoring software, permit a high degree of spacecraft autonomy. Many of the spacecraft now in production will operate with a significant amount of autonomy. This is especially useful for deep space missions, since the long cruise periods will then require fewer monitoring personnel. When it is not practical to place this capability on the spacecraft, new ground systems have been designed to reduce the operational workload.

Improvements in computer hardware and software also allow the spacecraft to achieve higher levels of processing on board. This reduces the downlink requirement and the burden placed on ground operators and equipment. For near-Earth missions, returning scientific data in near real time is an achievable goal. This enables researchers to use the spacecraft as an active instrument, instead of a passive data-gathering device.

The proliferation of high-speed communication networks, in conjunction with advances in autonomy for both spacecraft and ground systems, is allowing emphasis to be placed on science operations. It is now possible for the mission team, scientists and operators, to be geographically distributed. Scientists can remain at their home institutions while participating in near-real-time operations, reducing travel costs and improving mission effectiveness.

---

[12]Perspectives on the history of Wakeshield were provided by Dr. Michael Lembeck, Chief Engineer for the project, in a personal correspondence, March 19, 1997.

Commercial LEO communication satellites have accelerated the development of autonomous operating systems. Satellite constellations, such as Orbcomm, must maximize satellite availability and useful life. To accomplish this, designers have concentrated on automating repetitive tasks, such as the correction of simple failures and routine positioning. The commercial example illustrates that efforts to reduce operating costs should focus on automating routine activities, rather than "full autonomy" capabilities, for which the cost of development and implementation could outweigh the savings in reduced human labor (Tandler, 1996, p. 9).

Autonomy on the spacecraft can also increase mission risk. Pressure to reduce costs can lead to a premature commitment to autonomous operations and reductions in the number of ground operations needed to monitor spacecraft functions safely. It is important that the cost savings and performance improvements autonomous systems offer be carefully traded against the potential effects of failures on the spacecraft.

Spacecraft Standards and Commonality

Although no formal system of standardization has been applied to small spacecraft, ad hoc standards have emerged within the community. Reliance on MIL-STD-1553/1773 data bus standards has enabled many NASA programs to lower cost through reduced design load and the purchase and/or reuse of previous systems. Standard interfaces, such as the 1553/1773 bus, reduce integration costs and improve reliability because they usually reduce the number of electrical connections on the spacecraft. Many spacecraft are using 28 VDC power systems, allowing the expanded development of compatible equipment. Standards are equally important to spacecraft and ground operating systems. The continued evolution of standards is, therefore, very important in terms of reducing cost and increasing performance. Standards also encourage competition in the space-component marketplace.

Attendees at the RAND small spacecraft workshop recommended that the government not require the implementation of specific standards for small spacecraft but continue to observe voluntary standards. Voluntary standards, also called industry, nongovernmental, or consensus standards, are widely used within the space program. It should be noted that NASA has historically supported the development of voluntary standards, giving preference to them whenever possible. The Office of Safety and Mission Assurance (OSMA) is the organizational lead within the agency for preparing a coordinated response to emerging commercial, domestic, or international standards. (NASA HQ, 1991.) New standards are already evolving, and there are several initiatives within industry and trade organizations to foster additional refinement of hardware and software definitions. One concept was offered to NASA and DoD for possible consideration related to standards: creating incentives for small business providers of spacecraft components to help defray the additional costs of meeting an emerging standard.

THE DEVELOPMENT OF COMMERCIAL SYSTEMS

NASA's interest in IDIQ contracts and the possibility of procuring spacecraft under FAR Part 12 signals the arrival of an important commercial space milestone: the emergence of small spacecraft and their related systems as commodities.

Private-sector investment in the development of small spacecraft and related systems has been significant in the past five years. The price for standard commercial spacecraft buses is expected to be well below the cost of producing custom units. Commercial buses, such as the Lockheed Martin LM-700, offer the obvious advantage of reducing much of the nonrecurring costs associated with building spacecraft. Many firms are prepared to offer standard buses, and there appears to be a close match with many future requirements for small spacecraft missions. Plans for multiple spacecraft missions, such as observing constellations and formation-flying interferometers, seem particularly well suited to the purchase of commercial buses.

In the commercial bus model, technology infusion occurs in a stepwise fashion, with "next model year" buses integrating new designs and improved components. In addition to lower cost, commercial buses could offer the reliability implied in systems approaching mass-production status. Risk mitigation through "production quality" is an advantage, however, primarily when mission requirements match the performance criteria for which the bus was originally built. The LM-700 was built for Motorola's LEO Iridium communication satellite. It would require careful analysis to extrapolate its performance to other types of missions.

Commercial buses offer the greatest advantage in cases where the instrument and bus present a distinct interface, and little modification is needed to the basic spacecraft. As mentioned earlier in relation to the Air Force STP program, controlling requirements and forgoing modifications to the baseline spacecraft have proven difficult. Frequently, the instrument(s) represents a new design containing significant engineering uncertainty. Spacecraft and instrument design engineers communicate constantly during design and development. There is usually a good deal of give and take along the way, with a constant awareness that the instrument is the reason for the mission in the first place. It is not uncommon for the instrument to be "the tall pole in the tent" in terms of both cost and schedule.

Expanding private-sector offerings will likely also have a significant effect on the development of *mission unique* spacecraft, ones for which requirements are not amenable to purchase of a commercial bus. The proliferation of small communication satellites is expanding the available inventory of components and subsystems, many of which meet the requirements of science spacecraft. A commodity market in high-performance systems and components, such as GPS-based guidance packages and mass-memory devices, provides engineers with many more opportunities during the design process to "buy" as opposed to "make." Commercial systems are not always cost-effective, however. If reengineering is required to meet tighter performance specifications, the full cost of the commercial system can sometimes exceed the cost of a custom solution. Selection of commercial components requires careful trade studies.

THE IMPORTANCE OF PROCESS IMPROVEMENT

As opposed to advanced technology, which usually requires significant up-front investment, continuous improvement in the processes used to manage, build, and operate spacecraft promises roughly equivalent savings. One aspect of process improvement, initiatives to reduce the *Cost of Quality*, provides an illustrative example. Cost of Quality is the term of art for cost incurred, but rarely measured, in the execution of spacecraft missions; it is defined as:

> *Cost of Quality*—the cost of all efforts expended to find nonconforming output, to react to actual failures both before and after delivery, and to prevent failures from occurring in the first place. (Gilman, 1993, p. 2.)[13]

Cost of Quality is often equated with the cost of operating the quality assurance group. In practice, the concept is much broader, encompassing activities associated with failure prevention, appraisal, and recovery. At a workshop conducted at the University of Maryland in late 1993, NASA engineers estimated that the Cost of Quality ranged between 30 and 50 percent during the phases of a typical spacecraft development program. (Gilman, 1993.)

Not all Cost of Quality funds are recoverable. Prevention and appraisal costs will always be a component of mission costs. Creating a system that minimizes these costs, while remaining sufficiently robust to identify and correct deviations quickly, is the goal of Cost of Quality management.

Budget limitations will likely continue to be a primary well into the next century. Process improvement, applied to each aspect of spacecraft manufacture

---

[13]NASA's Office of Safety and Mission Assurance (Code Q) has sponsored additional cost-of-quality workshops at various field centers, the most recent of which occurred in August 1996. Infusing quality methods and metrics into engineering practices is covered in Kelada (1996).

to minimize loses in the system, is a key to continued cost reduction. Chapter Five contains additional discussion of process improvement, highlighting its importance in relation to risk reduction.

SOME HIDDEN COSTS

Spacecraft programs have worked hard to cut costs, but there are some downsides. To make small spacecraft missions *appear* cheaper, managers sometimes cross-fund programs, apply "excess" assets (from other, larger programs), and rely on excessive exchanges of government furnished equipment (GFE) (Creech, 1996, p. 3). On many occasions, costs are simply uncounted, a practice that can lead to the following:

- *Lessons not to be learned.* In small spacecraft programs, there is often little time or money to document team experiences. Travel funds are in short supply, discouraging the communication and cooperation required to vitalize new programs and train new people.

- *Uncounted risk costs.* Acceptance of risk has been a byproduct of cutting mission costs. The loss of a spacecraft like Mars Observer, built using cheaper, and riskier, technologies, is deeply felt (David, 1994). There are real costs associated with accepting risk, since presumably a lost spacecraft must be replaced. Risk acceptance might also be orthogonal to Federal policies moving in the direction of performance-based budgeting.

- *Poor working environments.* There is danger of creating "spacecraft sweatshops" with working conditions that exhaust and demoralize project personnel. Many of the projects examined in this study reported problems with employee fatigue, stress-related ailments, and retaining key staff. On flight projects, it is not uncommon to see employee time sheets in excess of 60 to 70 hours per week, much of it representing uncompensated time. To some degree, this is a fact of life in space development programs, but in past programs, such extremes occurred in the integration, test, and launch phases. The concern is that excessive workload is now appearing throughout the development cycle. Compounding this problem is the pressure to further trim schedules.

- *Loss of margin.* Most small spacecraft are being built with very small design and operating margins in an effort to save cost. Lean margins can drive *up* nonrecurring engineering costs, since it can be difficult to design systems to tighter specifications. Mission designers must also prepare and verify spacecraft operational sequences that have very little room for error. Finally, opportunities for commonality and standardization are frequently forgone because there is not enough money or design margin left to develop them.

- *Limited profitability.* Many commercial developers complain that small spacecraft are not profitable undertakings. Small spacecraft builders often operate as "skunk works" within a larger corporation. Their corporate viability can be tenuous in a low-profit environment. Capital equipment funds for tooling, new facilities, training, and certification are also hard to come by.

Although difficult to quantify, these costs are, nonetheless, real. Failure to account for them can affect long-term quality and performance.

ESTIMATING SAVINGS

To prepare a first-order estimate of the savings that have been produced by the various techniques outlined in this chapter, spacecraft program managers were asked to provide qualitative estimates. The results are shown Table 3.2.

The reasonableness of these estimates is tested by comparison to data for the XTE, in Table 3.3. Initial estimates of the projected cost of XTE were prepared by NASA GSFC's RAO (Strope, 1996, pp. 14 and 42). These estimates were based on historical GSFC spacecraft cost models. To trim costs, the Explorer program employed many of the innovative management, procurement, and technical approaches described in this chapter. The estimates from Table 3.2 would suggest savings for the mission of between 18 and 23 percent. The actual cost was 16 percent below the historical average. Some sources have estimated savings as high as 30 percent.[14]

Table 3.2

Cost Reductions in Small Satellite
Programs (percent)

Cost Element	Low Estimate	High Estimate
Management	20.0	30.0
Spacecraft Design & Development[a]	10.0	15.0
Integration & Test	20.0	25.0
Operations	15.0	30.0

[a]Includes spacecraft bus, instrument, software elements, and ground support equipment.

---

[14]"TRW Refutes Skeptics on Long-Term Future" (1995).

Table 3.3

Savings Applied to the XTE
Mission ($ millions)

Cost Element	XTE Projected Cost[a]	XTE Actual Cost	Cost Using Low Estimate	Cost Using High Estimate
Management	29.6	25.6	24	21
Space segment	153.7	145.8	139	131
Integration & Test	16.2	7.5	13	12
Operations	13.3	13.4	11	9
Total	229.2[b]	192.3	187	173
Percentage		16	18	23

[a]Based on historical data from NASA GSFC RAO.

[b]Includes a contingency of $16.4M.

SUMMARY

To bring down mission costs, NASA has adopted many changes in the way spacecraft are managed, designed, procured, and operated; the net effect has been savings of approximately 20 percent. These improvements affect all programs, large and small. There is reason to expect significant additional improvements as new technology and improved processes steadily evolve.

To achieve lower cost and better performance, NASA has relied more heavily on a maturing commercial sector for the building of spacecraft and on the science community for the management of missions. These changes underscore functional realignments that are occurring within the agency, changes that could affect core competencies. The desire for program cost-effectiveness should be balanced against a need to redefine, protect, and strengthen NASA's core competencies and the pursuit of mission excellence. Important in this regard is clarifying the criteria used to decide when spacecraft will be built in house (GSFC or JPL) and when they will be acquired from the private sector. Preserving the integrity of programs by ensuring that all costs are factored into mission planning is equally important.

Advancing technology holds great promise for increasing the performance of future small spacecraft programs. The role of new technology, and of issues related to planning and testing, will be covered in the next chapter.

Chapter Four

ADVANCED TECHNOLOGY FOR SMALL SPACECRAFT

To meet ambitious performance requirements, each new spacecraft depends to some degree on technological improvement. Incorporation of new technology has, however, traditionally been a cautious undertaking. Satellites have evolved in a stepwise fashion, with capability often lagging well behind the terrestrial state of the art.

A cautious approach to new technology does not necessarily indicate that mission managers are innately risk averse. Worth noting is the fact that science spacecraft, by their very nature, contain experimental technology: the instrument itself. Mission managers could perhaps be more accurately described as preferring to take only the required risk. The high success rate of past science missions indicates that risk avoidance worked, albeit at high cost.

Today, the pressure to maintain the pace of science while building smaller, less expensive spacecraft forces spacecraft developers to step beyond conservative boundaries. The mission manager must find ways to include new, and in many cases unproved, spacecraft and instrument technology amidst cost and schedule caps and demands for increased cost effectiveness. As mandated by NASA Headquarters, programs must also document and transfer new technology to the private sector.[1]

As a result, most of NASA's small science spacecraft incorporate an unprecedented amount of new technology. The incorporation of advanced technology is indeed a stated goal of some science programs. The AO for the Discovery program, for example, encourages the use of new technology. Candidate missions must identify new systems and components, analyze how the risks asso-

---

[1]The 1980 Stevenson-Wydler Technology Innovation Act required Federal laboratories to pursue technical cooperation with industry actively, while the 1986 Federal Technology Transfer Act made the transfer of technology the specific responsibility of all government research laboratories. An internal 1992 NASA review (Creedon, 1992) was highly critical of transfer practices, finding "little commitment from primary research organizations." The importance of technology transfer practices was subsequently emphasized in the The White House, Office of the Vice President (1993), which specifically directed that spacecraft missions draft technology transfer plans.

ciated with new designs are to be mitigated, and identify methods for transferring resultant technology within and outside of NASA.[2]

Technology plays a significant role in the cost-performance equation, and NASA clearly plans to rely on breakthrough systems to boost the performance of future missions. Some of the reviews conducted during this study suggest, however, that new methods for identifying, maturing, managing, and transferring spacecraft technology are as important a set of innovations as the technologies themselves.

This chapter examines several key aspects of NASA technology programs, focusing on their effects on smaller programs. The first section investigates how technology affects various types of scientific spacecraft. The next section reviews NASA investments in new spacecraft and instrument technology. The third section examines various sources of technology internal to, and external to, NASA. The fourth provides an overview of how technology is prepared for use on science spacecraft, including an analysis of the effectiveness of technology demonstrator missions. Technology distribution and utilization are covered in the fifth section. Each of these sections prepares a foundation for the final section, which focuses on the management and planning of NASA's technology initiative.

THE IMPORTANCE OF TECHNOLOGY IN FUTURE MISSIONS

Technology has a different focus depending upon the discipline being examined. This section studies how different science disciplines view requirements for new technology, based on interviews with mission managers.

Earth-science missions focus on advanced imaging instruments; bus upgrades are considered important secondary improvements. Small earth-observation missions, such as the SSTI Lewis and Clark, and NMP EO-1 spacecraft, primarily demonstrate new instrument technologies, while secondary developments aim to prove low-cost operational strategies and improved bus performance. Earth-science missions also characteristically maintain a well-defined bus-instrument interface, a feature that allows mission managers to consider the application of the expanding availability of commercial spacecraft buses. On a subjective scale, the focus might be considered as 70:30 in favor of an interest in instrument versus bus technology.

The space physics and astrophysical disciplines have a balanced (roughly 50:50) interest in bus and instrument technologies. These missions are highly opti-

---

[2]NASA, Office of Solar System Exploration Division, Solar System Exploration Subcommittee (1994), p. 15.

mized, and new technologies are sought to leverage performance on both sides of the bus-instrument interface. Spacecraft for observing astrophysical phenomena are also evolving along a path of evolution that can be termed "scalable." Spacecraft architectures are designed in such a way as to encourage reuse and to take advantage of lower-level advancements. These approaches could spill over to other classes of missions. Observatory spacecraft are also demonstrating higher levels of bus-instrument integration, a trend away from the application of standard spacecraft buses.

Planetary missions are completely dependent upon reliable spacecraft systems to deliver a suite of instruments to especially hostile destinations, suggesting a ratio of 30:70. Planetary spacecraft are highly integrated systems, and adopting standard spacecraft designs for planetary use has proven unsuccessful in the past. Planetary instrument developers are usually less concerned about squeezing the last drop of performance from a sensor or a measurement device than about preparing a system guaranteed to perform after a long journey into an uncertain environment. In an era of low-cost missions, planetary mission designers are especially dependent upon new technology to meet mission requirements. Technology development has focused on high-efficiency power systems, microelectromechanical devices, autonomous systems, and miniaturized instruments. The importance of technology to planetary missions is reflected in the advanced technology development (ATD) budgets. In FY 1996, ATD spending for planetary missions ($15.9 million in FY96) was double that of astrophysics ($8 million) and quadruple that of space physics ($3.7 million). (OMB, 1996.)

LEVELS OF INVESTMENT IN SPACECRAFT TECHNOLOGY

Advanced technology is vitally important to the future of small spacecraft. Small spacecraft have demonstrated an ability to return first-class science, but to meet the expanding expectations of the space science community, more performance is needed. If higher levels of performance and lower mission costs are to come from small spacecraft, adequate investment in technology development must be assured. Further, advanced technology funds must be carefully mapped to realistic performance expectations.

A review of NASA mission-level plans, technology road maps, and ATD programs presents an array of projects that are not clearly linked to a strategic investment plan. Additionally, establishing the level of NASA's investment in technology is complex because many projects lie entangled within mission developments or seemingly unrelated line items in NASA's budget.

OSS's R&A budget, for example, contains ATD projects related to advanced instrument technology. In FY96, ATD accounted for approximately 13 percent of OSS's $210 million R&A budget (OSS, Solar Physics Division, 1996). The mixing of technology development with scientific investigation can also be seen in Solar Physics Supporting Research and Technology (SR&T) program. Of the 37 proposals selected following the NASA Research Announcement (NRA) in May of 1996, five supported advanced technology development for future instruments (OSS, Solar Physics Division, 1996).[3]

Some technology-related lines in the NASA budget contain components that arguably could be considered unrelated to development of specific technologies. NMP, for example, is an advanced technology development effort, but the majority of these funds are dedicated to building demonstrator spacecraft. An estimated 28 percent of the NMP Deep Space One TMC is dedicated to the development of new technology; the rest goes to developing the spacecraft (NMP Office, 1996).

Table 4.1 presents an estimate of technology investment based on available data from NASA's FY96 budget and from analyses of individual programs. This estimate includes clearly identified technology programs within OSS, Mission to Planet Earth (MTPE), and the former Office of Space and Advanced Technology (OSAT); relevant percentages of the aforementioned R&A and SR&T accounts; and the technology contributions of programs, such as NMP. The estimate excludes advanced mission definition (Phase A/B) spending, because it was judged to be unrelated to technology development. Table 4.1 estimates a total spending of $249 million. This represents just over 6 percent of the budget of the three offices, or less than 2 percent of the $13.9 billion FY96 NASA budget.[4]

The next section will examine the sources of spacecraft technology. NASA small spacecraft are heavily populated with systems from commercial and DoD sources, suggesting that NASA investments represent a small contribution to the development of advanced systems. It is important to consider, however, that NASA's level of investment may be adequate if the agency's mission is to focus on the production of technological solutions unique to the requirements of its stakeholders. The Solar System Exploration Subcommittee recommended that spending on advanced technology for planetary missions attain a level of 5

---

[3]R&A and SR&T accounts usually contain ATD and MO&DA elements. NASA's Solar System Exploration Subcommittee in September of 1994 noted that these accounts "have been used as a reservoir of funds to fix problems . . . for which funds were not allocated in the budget process." NASA is experimenting with new budgeting schemes in which these cost elements are separately accounted for.

[4]A similar estimate of NASA technology spending can be found in Augustine Committee (1990), p. 31. Investment in spacecraft technology was cut by more than half from the level of the early 70s, but has been relatively steady at approximately 2 percent of NASA's budget since 1975.

Table 4.1

NASA's Space Technology Investment

Office	FY96 Budget ($B)	Estimated Technology ($M)	Percentage of Budget
OSS	2.03	48	2.3
MTPE	1.29	27	2.1
OSAT	0.62	174	28.1
Total	3.94	249	6.3

to 7 percent of the Solar System Exploration Division's budget, a figure in keeping with the 6 percent of the combined OSSA-MPTE-OSAT budget in FY96 (NASA Solar System Exploration Subcommittee, 1994).

SOURCES OF SPACECRAFT TECHNOLOGY

NASA's investment in human spaceflight and large missions "has resulted in minimal spending for advanced research and development" for small spacecraft (NRC, 1994, p. 7). Within the past decade, as the demand for small spacecraft has increased, developers within NASA have relied heavily on DoD and commercial sources of technology. Radiation-hardened computer processors from Air Force and BMDO programs can be found in abundance inside of small science spacecraft. The aforementioned WIRE spacecraft is using a solid hydrogen cooler demonstrated on previous military spacecraft. There is some concern within the spacecraft development community that the technological windfall from SDIO/BDMO and Air Force investments in the late 80s and early 90s will evaporate in the wake of sharp DoD cutbacks.

Although outside sources have supplied key technologies that have allowed NASA to develop increasingly sophisticated small spacecraft, the agency's technology investment does support an extensive research and development (R&D) capability. Many individual NASA spacecraft programs have taken steps to improve their internal technology management practices. The Explorer Program, for example, has created a clearly identified line item within its budget dedicated to small spacecraft technology development.[5] Important develop-

---

[5]Many individual technology development efforts are under way within OSS and MTPE. In some cases, such as the Mars Instrument budget item, the programs are highly visible. In other cases, such as the Planetary Instrument Definition and Development Program (PIDDP), a specific initiative is nested within a larger budget element and is, therefore, less visible.

ments at the working level, however, cannot replace the need for an effective NASA-wide technology planning and management structure.[6]

Reflecting the agency's dual mandate to conduct basic research and apply advanced technology, NASA has traditionally maintained two technology cultures:

- *Research-oriented groups*—organizations that are closely associated with basic and applied R&D programs. For example, at LeRC the Photovoltaic Branch of the Power Technology Division "is dedicated to the advancement of solar technology for the purposes of meeting NASA's and the country's energy needs." (NASA Lewis Research Center, 1995.)

- *Mission-oriented communities*—organizations closely connected to flight programs, such as the Space Technology Division at GSFC. This group is responsible for "executing tasks assigned by the flight projects" in relation to "the design, development, and management of spacecraft and instrument components . . . ." (NASA Goddard Space Flight Center, 1996.)

Mission-oriented development groups, closely aligned with flight teams, have proven highly successful when it comes to meeting technology needs, both at GSFC and JPL. They provide the necessary responsiveness, enjoy proximity to their customer base (often a short walk) and are loyal to the mission team. They provide "homespun" technology solutions tailored to mission requirements. Dependence on the local technology store, however, can lead to limited choice and an insular approach to building spacecraft.

Research-oriented groups, on the other hand, have traditionally been funded from NASA's technology budget.[7] They are not connected directly to flight programs; rather, they surround the research facilities that support their basic research focus. Research-oriented groups have spurred the development of electric propulsion, low-gravity fluid-handling equipment, high-power generation systems, advanced structures, and a host of other innovations.

These two cultures are rooted at opposite ends of the technology-development spectrum. Occasionally, they meet in the middle; but frequently, they do not. Budget reductions have invited these two communities to communicate more effectively. A variety of conferences related to small spacecraft technology have

---

[6]Several national reports have focused attention on NASA's traditional lack of effective technology management practices. The most notable are NRC (1987), Augustine Committee (1990), NASA Space Systems and Technology Advisory Committee (1991), and NRC (1993).

[7]Research-oriented groups were traditionally funded by the recently abolished Office of Space Access and Technology (OSAT) or, before that, the Office of Aeronautics and Space Technology (OAST). Mission-oriented groups were generally funded from within the science budgets.

helped to bring technologists, spacecraft builders, and scientists into closer contact. RAND workshop participants noted that the partnership model created by the New Millennium Program was particularly effective in engendering communication between the various communities.

Research and mission-oriented groups both contribute to an overall process of creating new technology. To ensure rapid infusion into flight programs, however, new designs must mature in a way that is both cost-effective and responsive to the needs of the end users: the spacecraft builders and mission scientists.

APPROACHES TO MATURING SPACECRAFT TECHNOLOGY

An advanced technology, whether it is a part, component, assembly, subsystem, or turnkey system, is useful to a spacecraft developer when it has matured to the point that it can be incorporated into a flight project with acceptable engineering, cost, and schedule risks.

There are many ways to reduce the risk of using new technology in science missions. The most complete way, and the most expensive, is to demonstrate the device or system in space before its intended use. In an era of constrained budgets, however, all methods must be explored and the most cost-effective means must be fully exploited. Efforts to mature new technology must also take into account the fact that technology has not been a major source of failure in past missions (Kicza et al., 1997, p. 4). The methods selected to reduce risk should also be tailored to small spacecraft applications. Here, a review of what constitutes "flight certification" is valuable.

Defining Flight Certification

The maturation of space technology is captured in NASA's Technology Readiness Levels (TRLs), as depicted in Table 4.2. TRLs are important to the management of technology programs and are essentially a step-by-step risk schema for *retiring* risk. According to NASA plans, funding for advance technology shifts from the *developer* to *user* as a concept moves to a higher level of readiness. The user, a science mission for example, would expend only limited resources in Levels 1–4 (mainly to identify and track requirements); begin to pay an increasing share during the demonstration phases, Levels 5–7; and then completely fund the flight phases, Levels 8 and 9.

TRLs imply formality in the process of maturing technology. Rigid application would mean, however, that a system could not be considered "flight qualified" (TRL-8) unless an earlier prototype (TRL-7) had flown in space. In practice, few development efforts move sequentially along the TRL continuum.

Table 4.2

NASA Space Technology Readiness Levels

Stage[a]	Level[a]	Objective[a]	Funding[b]
Basic Technology	1	Observe and report basic principles	Developer
Feasibility Research	2	Formulate concept or application	Developer
	3	Prove the concept through analysis or experiment	Developer
Technology Development	4	Validate the concept using components or breadboards in the laboratory	Developer
Technology Demonstration	5	Validate the component or breadboards in a relevant environment	Cofunded
	6	Demonstrate a model or prototype in the relevant environment, ground or space	Cofunded
System/Subsystem Development	7	Demonstrate a prototype in the space environment	Cofunded
System Test, Launch and Operations	8	"Flight qualification" of the actual system through ground or space test	OSS, MTPE
	9	"Flight proven" through successful mission operations	OSS, MTPE

[a]OAST (1991).

[b]OSS (1995a).

Terms like *flight certified* and *flight qualified* bear close scrutiny. These terms are amalgams that embody concepts related to reliability, traceability, and quality. When applied to new technology, these terms imply that a certain standard has been achieved and that performance assurance will follow. In practice, there is reason to question an unfettered application of these terms, especially in relation to small, low-cost spacecraft.

Rather than a demonstrated engineering relationship to risk, *flight qualified* is a term that carries only a *perceived* standard of quality or readiness for use. The definitions outlined in Table 4.2 are not mapped to specific inspection or quality assurance standards, test requirements, or material specifications. Even if such a correlation existed, "flight qualified" technologies are not assured to work in a given application. Also, the definitions used within NASA are not universal; DoD and commercial manufacturers have different nomenclatures.

The *flight proven* label is also deceptive in that it can create a false sense of security about using a design or system. Equipment that has a prior flight *heritage* might not be applicable to another mission design. Application through inheritance has been used in the past to select components and devices, often with insufficient testing. This strategy is thought to be one of the main causes of the failure of the Mars Observer spacecraft.[8] Reliance on labels such as

---

[8]The Mars Observer spacecraft suddenly stopped transmitting on August 21, 1993. Without diagnostic telemetry, the cause of the accident will never be known. The failure investigation board did

"flight qualified" and "flight proven" without careful review of applicability can introduce unanticipated risk.

Most science spacecraft are built with some designs that have not yet been tested in space. At the part and component level, both NASA and DoD maintain preapproved lists of parts that can be selected and applied without extensive testing. The use of "nonstandard" parts (items that have not been preapproved) is, however, increasing, most notably the growing use of commercial electronic components. In the case of these items, procedures exist to establish whether or not the part is likely to survive operation in the space environment.[9] At a higher level of integration (for example, a new type of attitude control subsystem), NASA and military handbooks and engineering specifications specify the design and test procedures necessary to access the applicability and associated risks. There are, therefore, well established methods for assessing a new technology and gaining confidence in its performance. Yet, the ability to evaluate a new technology does not ensure its acceptance. New technology must be oriented to the requirements of the end user and will be accepted when a benefit is perceived (Creedon, 1992, p. 6). Discussions with spacecraft builders produced a notional set of criteria that a new technology should meet:

- demonstrate repeatable performance in conditions similar to those expected aboard the spacecraft

- adequately assess all risk factors

- employ high quality components with lineage to known standards or to test data that establish reliability

- be sufficiently supported (development software, integration and test procedures, parts, etc.)

- promise clear performance gains over existing technology

- present a cost commensurate with performance

- have interfaces that are documented and that can be configured to match other systems.

Since cost reduction is of central importance to future missions, a new, and hopefully universal, understanding may be needed of what it takes to *retire risk*

---

note, however, that reliance on equipment and designs from LEO satellites likely contributed to the loss of the spacecraft. Similar problems were encountered on the successful Magellan mission to Venus. During the mission, several problems were encountered with equipment that was inherited from earth-orbiting spacecraft, such as the star scanner and tape recorder. NASA engineers later concluded that additional ground testing would likely have revealed many of these problems; see Gonzalez (1996a), p. 20.

[9] For example, see GSFC (n.d., a).

adequately in systems to be flown on small spacecraft. One possible means of establishing this understanding is to bring technologists and spacecraft designers together expressly to create mutual definitions and maturation criteria. The product assurance (PA) function could perhaps serve to assist technology projects to mature in such a manner as to ensure meeting the requirements of spacecraft projects.[10]

Low-Cost Means of Maturing Technology

Usually, new designs can be adequately matured using low-cost, ground-based environmental simulators, such as drop towers, low-g simulator aircraft, and space thermal-vacuum chambers. An NRC report (NRC, 1994) recommended that technology for space physics applications be advanced by utilizing suborbital platforms, such as rockets and long-duration balloons, and unique ground sites, such as the polar cap. The recent success of the Mars Pathfinder mission is clear evidence that ground-based opportunities are usually adequate to test new designs successfully.

There are times, however, when a new technology must be flown in space; deployable systems are an example. Precision apertures, tethers, and inflatable space structures are difficult to test on the ground and they represent considerable technical risk. There are a variety of low-cost methods of testing these systems. The NRC recommended placing new technology on planetary, DoD, and other spacecraft of opportunity, and flying precursor units for flight certification on earlier missions. The use of small spacecraft as precursor instrument platforms for larger missions follows NASA tradition. For example, the Clouds and Earth Radiant Energy System (CERES) instrument, scheduled to fly on the MTPE's AM-1 spacecraft, will fly earlier on TRMM. The recent flight of the Naval Research Laboratory's Tether Physics and Survivability Experiment (TiPS) is an example of launching technology payloads on a host spacecraft (Alfriend et al., 1995, p. 2-5). TiPS was a simple, very low-cost demonstrator that provided engineering data to validate complex models of the structural response of tethered satellites. Another example is the New Millennium Deep Space 2 experiment, an instrumented penetrator, that is being carried to Mars on the 1998 Mars Surveyor Lander (Gavit, 1997). In addition to the options listed above, new technology can be tested using

---

[10]Here, the term "product assurance" is being used in the broadest sense to include reliability and quality assurance (R&QA) and mission assurance (MA), even though in many circumstances these terms describe distinct activities and sometimes different organizations. At NASA, PA activities are coordinated by the OSMA.

- *Secondary strings.* New components, devices, and subsystems can be flown as backup or redundant elements of primary systems on mission spacecraft.[11]

- *End-of-life testing.* A phase at the end of a spacecraft mission can be reserved for evaluating new technology. One such program, the Flight Test Bed for Innovative Mission Operations (FTB-IMO), is dedicated to testing operational concepts (Bruner, 1995).

- *Class D payloads.* Fast-turnaround systems, such as the Shuttle-based Spartan carrier, Get-Away Specials (GAS), and the Hitchhiker Payloads (which can deploy microsatellites from the Shuttle cargo bay), offer opportunities for testing new spacecraft and instrument technology at low cost.[12]

The recent Inflatable Antenna Experiment (IAE) illustrates the exploitation of Class D payload carriers for low-cost technology demonstration. Figure 4.1 shows the IAE deployed from its Spartan carrier after release from the Space Shuttle. The relatively low-cost Spartan was a good match for the risks associated with the demonstration, which was only partially successful.

Use of Dedicated Flight Demonstrator Missions

The most complete way of retiring risks associated with new technology is to employ dedicated flight demonstrator spacecraft. As pointed out earlier, new technology can be found on virtually every one of NASA's science missions. To some degree, each mission has technology demonstration objectives. A dedicated flight demonstrator, however, is a standalone mission in which a non-reusable spacecraft is built expressly to test multiple technologies simultaneously. Examples of such missions are the SSTI Lewis and Clark and NMP Deep Space One and Earth Observer One spacecraft. The Discovery Mars Pathfinder mission was also considered a demonstrator mission to test planetary micro-rover technology and other techniques. Each demonstrator has a scientific component, but the primary mission is to verify the performance of advanced

---

[11]A redundant string is a secondary path in a spacecraft system. New designs are often first flown in a redundant environment, sometimes as the primary system with a traditional design in a secondary role, or vice versa. Another option for testing new designs is to dedicate a portion of a system for new technology. New types of solar cells, for example, are flown in a segment of the array, with more traditional cells in the primary area. The performance of the new cells can be evaluated using this technique with very little risk to the performance of the spacecraft.

[12]The Spartan free-flyer operates routinely. Details of the IAE mission can be found in GSFC (1997). In addition to the IAE, a Spartan was used to test commercial plastic-encapsulated electronics, a controversial application that is discussed in more detail in Appendix D; see Garrison (1996).

Figure 4.1—IAE Erection on STS-77

systems and concepts. For example, on New Millennium Program missions, 90 percent of the objectives relate to obtaining validation data on new technologies. (Ridenoure, 1996, p. 4.)

DoD also relies on demonstrator spacecraft, but NASA and DoD have different objectives. Military demonstrator missions, such as the long-lived STP initiative, are proof-of-concept flights, essential precursors to the large investments to be made in deploying *operational* networks. In military systems, "the ATD serves as a stepping stone for follow-on operational systems, lowering critical risk and cost uncertainties." (Worden, 1994.) The national security aspect of the future asset further underscores the importance of precursors. Although often remembered as a lunar science mission, the principal objective of the Clementine spacecraft was the testing of future operational sensors and systems for the Brilliant Eyes and Brilliant Pebbles programs.

NASA can claim neither national security nor future operational status as a driving requirement for demonstrator spacecraft. NASA has also claimed a willingness to accept more risk in small spacecraft programs. In lieu of these considerations, it must be assumed that NASA demonstrator missions represent especially high-risk technologies.

NASA demonstrator spacecraft introduce a new objective: *flight validation*. The objective of NMP, for example, is "validating key technologies that can significantly contribute to lowering life cycle costs and increasing scientific return." (NASA Headquarters, 1995.) It is not readily apparent how a *validated* system is related to a *certified* or *proven* system, or to TRL definitions of maturity. NASA demonstrators also seek to validate a suite of high-risk technologies simultaneously; this is an ambitious undertaking. (David, 1997b, p. 2.)

The underlying strategy of NASA's demonstrator program is one of revolution versus evolution. The driving requirement is the shift to the higher levels of performance needed to meet future science requirements—to reach the up-sloping new performance curve shown in Figure 2.4. This is an important objective, but one that can be quite difficult to attain in practice. Some of the factors that can complicate the development of dedicated demonstrator spacecraft are:

- *Cost.* A demonstrator spacecraft can meet or exceed the cost of a science mission—finding the money to fund such spacecraft is challenging. In FY96, NASA demonstrator missions constituted an annual investment of over $300 million. Attempting to integrate several technologies exposes the mission to significant cost, schedule, and technical risks.

- *Risk Elimination.* Under pressure to maintain a flight schedule, higher-risk technologies might be bumped from the flight or placed in redundant strings. Placing new technology in redundant strings weakens the justification for the demonstrator mission, since the new design could presumably perform the same function on a science spacecraft.

- *Risk Neutralization.* The mission manager, usually constrained in terms of cost and schedule, must select a compatible set of reasonably mature technologies when designing the spacecraft.[13] Choosing relatively mature designs, however, further weakens the justification of the demonstrator, since design maturity might be sufficient for a science spacecraft.

- *Performance Discrimination.* In the event of a partial failure, it is often quite difficult to ascertain which systems performed adequately. A catastrophic loss can mean the failure to "validate" any of the candidate technologies.

- *Delayed Availability.* Presumably, a new technology will not be available until the demonstrator flight is complete, delaying its use for 2 to 3 years.

An additional goal of this type of mission is the demonstration of new technology in environments similar to what is expected on subsequent science missions (Ridenoure, 1996, p. 3). This goal is somewhat suppressed by the increas-

---

[13]The NMP EO1 spacecraft is being built under a fixed-price contract. The DS1 spacecraft also operated under cost constraints; see Lehman (1996), p. 32.

ing fidelity of space environmental models, which should be able to predict the performance of new technology with greater assurance. It is also unclear how well a successful performance in one environment can be extrapolated to others.

Since funds for demonstrator missions must be carved from a tightly constrained budget, it seems important that they be used selectively and with great effect. Justification of these missions should be exceptionally thorough, and performance measurements equally rigorous. One possible method of ensuring that dedicated technology demonstrators are adequately justified is to ensure that the risks being addressed are real and not perceived. This can be accomplished by establishing a clear link between past failures and proposed technology plans. Peer review, with oversight from the spacecraft development community, might also be considered as a means of ensuring that demonstrator missions have met appropriate tests and criteria.

TECHNOLOGY DISTRIBUTION AND UTILIZATION

NASA spent approximately $30 million on technology utilization activities in FY96. These efforts mainly focused on the *external* use of NASA-developed space technology. Internal technology transfer is, however, equally important.

Many participants at the RAND workshop reported that technology distribution has been hampered by reductions in program budgets. In lean programs, it can be challenging to divert human resources from design and test functions to the packaging and reporting of new developments. The personnel most familiar with a new design concept are often in the highest demand throughout the flight program and beyond, further compounding the job of documentation.

Reduction in the amount of formal documentation required from programs can lead to a decrease in the information and categorization of new designs and product developments. Some cost reduction practices, such as the acceptance of drawing redlines or the reduction of software documentation, can make design reuse difficult and in many cases impossible.[14]

Incentives for Technology Transfer

In response to external and internal recommendations, NRAs, AOs, and Requests for Proposals (RFPs) will carry technology-transfer requirements.

---

[14]The practice of accepting "redlines" (handwritten engineering notations) on production drawings helps reduce cost by allowing engineers to make on-the-spot changes during fabrication. Often, when the spacecraft has been completed, insufficient funds remain to update drawings to reflect the "as-flown" configuration.

Technology transfer does not come without cost to programs. RAND workshop participants suggested that NASA could incentivize technology transfer by offering technology investment awards up to some percentage of a program's TMC. These awards would be available to the principal developer of the spacecraft, government or commercial, to recover costs associated with maturing new designs to a level that they become turnkey systems to other programs.

The Internet has also assisted smaller teams in transferring technologies and design practices. Most spacecraft developers maintain a high level of proficiency in the use of computer information services to help meet the requirements of completing complex programs. The proliferation of the WWW has enabled spacecraft engineers to share design information in ways not previously possible. Nearly every small mission has an associated web page, which provides information related to the techniques and designs used to implement a given mission.

Databases have been developed to capture spacecraft technology. They have been of limited utility in the past, partly because development efforts did not enjoy sufficient high-level endorsement. Broader use of databases, especially of systems accessible through the Internet via WWW servers, could greatly assist the sharing of spacecraft design information.

PLANNING AND MANAGEMENT OF SPACECRAFT TECHNOLOGY PROGRAMS

At no time in NASA's history has the need to coordinate, justify, and manage spacecraft technology programs been greater. As the importance of technology to space science missions has increased, so has the frustration that NASA's approach to developing and integrating new technology has been inadequate. The 1996 abolition of OSAT at NASA Headquarters reflects the perspective of senior agency managers that, with limited budgets, the agency can ill afford technology development not directly connected to flight programs. ATD programs have been labeled "hot dog stands," disconnected from the mission base.[15]

Overall Technology Planning

The need for an overall NASA agency technology plan was highlighted in the 1990 Augustine Committee report. The report recommended "that an agency-

---

[15]"Demise of NASA Advanced Technology Unit Viewed with Alarm" (1996).

wide technology plan be developed with inputs from the Associate Administrators responsible for the major development programs." (NASA Advisory Committee on the Future of the U.S. Space Program, 1990, p. 31.) In response, OAST issued the Integrated Technology Plan (ITP) for the Civil Space Program. A 1993 NRC report *Improving Technology for Space Science* was a detailed review of this plan. The central weakness of the ITP was that many of the elements of a plan were missing: a clear methodology for managing technology development, tying technology to space missions, and evaluating success. The 1993 NRC review concluded that the ITP was essentially

> a prospectus of development tasks most of which cannot be undertaken within either the existing budget or any budget that is likely to be available. . . . (NRC, 1993, p. 4.)

Following the 1993 NRC review, progress in preparing a new top-down technology plan within NASA awaited the creation of OSS's Integrated Technology Strategy. (OSS, 1995.) The strategy was developed to correct specific weaknesses NASA identified in the prior plan, namely:

* No integrated means of identifying, developing, and inserting technologies into the system

* Insufficient attention to life-cycle costs

* No overarching technology development strategy

* Insufficient resources directed to technology development requirements

* No formal criteria for technology selection, funding, or transition planning

* The tendency for flight projects to shun new technology because of cost, schedule, and risk implications

* Lack of a robust technology definition and development process

* Ignorance at the project level of technology transfer imperatives

* No prescribed metrics for effectively evaluating technology transfer success.

Yet lack of a top-down plan has not prevented "grass roots" planning. The need to cut costs while increasing spacecraft performance leaves mission managers with little choice but to prepare detailed, budgeted plans for technology infusion. Mission-level plans have improved dramatically in the last two years. Now most of NASA's small spacecraft flight programs (Explorer, Discovery, etc.) have refined technology plans that are being executed with increasing precision.

Today, the need to push spacecraft as close to the state of the art as possible is generating a concept known as "just-in-time technology." This strategy aims to deliver the highest possible level of performance into a new design without incurring costly delays. That managers believe this to be possible is an indication of the sea change taking place inside of spacecraft design teams.

Recent Technology Planning Actions

On September 4, 1996, NASA announced plans to create the Office of Technology (OT), housed within the Office of the Administrator. OT is a small, coordinating organization to oversee the agency's technology budget. It also reviews technology programs and distributes them to the Aeronautics, Human Exploration, MTPE, and Space Science Enterprises (Mulville, 1996).[16]

Both MTPE and OSS fund the development of spacecraft technology. When OT was created, OSS was given a lead responsibility in that it is charged with planning technology programs that are common to both space and earth science missions. Within OSS, the Advanced Technology and Mission Studies (AT&MS) Division was formed to perform both cross-cutting technology planning and initiatives unique to space science.

A principal objective of OT is to reinvent the ITP. In helping to prepare an updated version of the ITP, AT&MS has classified individual technology projects according to breadth and maturity. Technologies that are narrowly focused, that support missions in either the Space Science or Mission to Planet Earth enterprises, are distinguished from those that potentially have broader appeal. Near- and mid-term technologies are also distinguished from ones likely to mature only in the far term. Of the available funding, the split is slated to be on the order of 75 percent to near- and middle-term technologies and 25 percent to far-term technologies. To match mission requirements to individual technology initiatives, Joint Planning Teams have been formed in key disciplines (telerobotics, communications, instrument sensors, etc.). Coordination among teams and across enterprises is to be ensured through a Joint Enterprise Strategy Team (Ulrich, 1997).

The creation of this plan is only now maturing, and it is too early to evaluate the success of proposed strategies. Review of an early strategy, however, suggests some areas on which to focus attention:

- *Plan review.* Although NASA's new technology-planning process recognizes the importance of external reviews, it is not clear how these reviews

---

[16]NASA is organized into four strategic enterprises that function as primary business areas for implementing the agency's long-range vision. In addition to the Space Science and Mission to Planet Earth enterprises, NASA also operates the Aeronautics and Space Transportation Technology and Human Exploration and Development of Space enterprises.

will be conducted and how review decisions will affect the funding of technology projects. Peer reviews are recognized as "sensible" only up to the TRL 3 level (presumably because technology projects above this level might contain information considered proprietary).

- *Planning elements.* It is not clear how this current effort will escape criticisms of earlier plans—namely, that elements of a plan (budgets, schedule, milestones) were missing.

- *Exit criteria.* Clear definitions are not yet available that identify how and when technology programs will be terminated. Likewise, the connection between future scientific mission requirements, technology road maps, and performance metrics for individual projects is unclear.

- *Coordination.* Mechanisms for sharing information with other government and private-sector spacecraft technology organizations have not yet been proposed.

- *Planning base.* The extent of the effort is not clear. In relation to spacecraft technologies, for example, it is not evident whether the ITP will extend beyond spacecraft bus technologies to the development of instruments and sensors, ground systems, design processes, and test facilities. The authority of this plan over the mission-level plans and budgets of NMP and the Explorer Program's technology initiatives is also difficult to ascertain.

The AS&MS effort to create a new NASA ITP is clearly distinguished, however, in that it establishes a contract relationship between NASA Headquarters, the technology developer, and the end user. The goal of creating this cooperative relationship is critically important to small spacecraft programs that must increasingly rely on advanced technology.

SUMMARY

One of NASA's most important challenges is to advance small spacecraft performance and reduce mission cost by incorporating advanced technology. National space objectives increasingly rely on successful small missions, which, in turn, rely more heavily on higher-performance systems and components. An effective means of planning and implementing an aggressive technology program is, therefore, essential.

Past reviews have criticized NASA's technology planning efforts. In response, the agency produced the 1992 ITP for the Civil Space Program. While the ITP attempted to centralize technology planning, many elements of a plan were missing: a clear methodology for managing technology development, tying technology to space missions, and evaluating success. In 1996, NASA disbanded the OSAT and placed responsibility for planning all spacecraft technology programs within the new OT. A central objective of this realignment was to refine and reissue the ITP. The success of this plan is a matter of some import to managers of small programs tasked with integrating the advanced systems needed to meet ambitious scientific milestones.

The task of preparing an integrated plan is complicated by many factors. Perhaps the most important challenge NASA faces is the fact that the agency has traditionally had a dual mandate: to conduct basic research *and* to develop applications for new technology. This duality has spawned separate cultures within the agency. *Mission-oriented* groups are closely aligned with, and responsive to, in-house spacecraft builders. These groups supply important incremental advances. *Research-oriented* groups have not been associated directly with flight programs unless a specific technology was being tested. Their basic research orientation has supplied some of the more revolutionary advances. NASA's technology planning effort must unite these cultures and engender cooperation to an unprecedented degree. Another factor is the insularity associated with reviews of NASA technology projects in the past. Technology projects were reviewed, but the review process itself could be characterized as internalized. The 1990 Augustine Committee recommended that technology funds be allocated based on a review by experts outside of NASA. RAND workshop participants were unanimous in their agreement with this recommendation. To support such a review, the ITP must contain cost, schedule, and performance data for the many technology projects that NASA sponsors.

Creating the ITP requires that NASA conjoin several other technology planning efforts. Technology planning occurs within several program offices and at various field center locations. To claim full integration, the ITP must assimilate or coordinate these efforts, in turn implying a high degree of trust between those building spacecraft and those developing technology. NASA has recently created an internal contract structure to ensure that technology, planned within the framework of the ITP, will be ready when needed. For this contract to work, an assurance is needed that the methods and terminology technologists use to measure and describe the readiness of new designs are relevant to the spacecraft builder. Current practices do not offer this assurance.

The ultimate output of the ITP will be advanced designs that are ready for use on future science missions. NASA currently relies heavily on dedicated technology demonstrator missions, like SSTI and NMP, to ensure that new systems are ready for use. In FY96, nearly one-third of NASA's $1 billion investment in small missions was used to construct technology-demonstrator spacecraft. These missions are expensive and contain significant technical and programmatic risks.

In an era of tightly constrained budgets, it is important to carefully determine when dedicated technology missions are the most cost-effective means of retiring technological risk. The performance of new designs can usually be adequately evaluated using low-cost, ground-based approaches. When a new technology must be tested in space, additional low-cost test methods are available. These methods have proven successful, and new technology has traditionally not been the source of mission failure.

The cost of dedicated technology-demonstrator spacecraft should also be balanced against potential gains from alternative investments, such as expanding the base of fundamental research and technology programs. Further, in establishing this balance, it is important that the risks these missions are designed to retire be real and not simply perceived. A possible means of achieving this would be to generate the requirements for these missions from within NASA's ITP, which, in turn, should base the assessment of technological risk upon actual flight failure and performance data.

Chapter Five

PROCESS IMPROVEMENT AND RISK MANAGEMENT

This chapter links two important concepts: process improvement and the management of risk. Each concept is important in its own right. Process improvement strategies provide a means of meeting future cost and performance goals. Similarly, NASA's space- and earth-science missions are expressions of national space policy, and the risks associated with them need to be understood. In relation to NASA's programs, however, these concepts are more closely linked. This linkage occurs at the point of *risk measurement.*

Communication of risk desires and outcomes between policymakers and NASA cannot occur unless risk can be expressed, either qualitatively or quantitatively. Likewise, process improvement theory is based on the principle of measuring all product variables: cost, performance, *and* reliability (or, in a related manner, risk). NASA has not, however, formally identified risk reduction as a strategic goal; process improvement initiatives can, therefore, be assumed incomplete.[1] The agency is, however, being driven at the technical level to refine methods for calculating risk. New quantitative risk measurement methods are needed mainly because NASA can no longer afford to avoid risk at any cost and must manage it as an engineering variable. Efforts to refine risk measurement methods at the technical level could supply (a) an important metric for overall process improvement measurement and (b) a mechanism for reporting risk to the policymaker.

This chapter will review the concepts of process improvement and risk management and will more firmly establish the link between them. The first section discusses the relevance of these issues to the policymaker, while the second describes the key attributes of process improvement and its importance in other industries. The third reviews new approaches to risk management and how they, along with cost and performance measures, support process

---

[1]RAND notes that the July 3, 1997 draft version of the OSS Strategic Plan now includes risk reduction as a long-range technology goal.

improvement and the communication of risk. The penultimate section examines four key trends that could lead to the desired cost, performance, and risk outcomes. The chapter concludes with some suggested strategies for improving processes and reducing risk.

RELEVANCE TO THE POLICYMAKER

Currently, there is no systematic method for NASA to communicate risks to the policymaker. Discussions of risk are limited to larger programs when perceived failure costs are high. Yet, issues related to risk have implications far beyond program-level boundaries:

- *National prestige.* International agreements are often at stake, and failures can have unexpected repercussions.[2] Cumulative failures, even small ones, can seriously jeopardize national space goals and cause a loss of momentum.

- *Health and safety.* For some spacecraft, environmental impact and the public safety become major issues, as in the case of flying radioactive thermal generator (RTG) power sources. In these cases, risk becomes a high-profile national issue requiring approval of the President's Science Advisor.

- *Response to policy.* Program-level decisions regarding risk may not correctly interpret the intention of Federal laws, such as the GPRA. Policy guidance and interaction are required to inform implementing organizations of the goals and intentions of rule-making.

Risk measurement is also important in terms of overall performance measurement. A recent NRC report noted that

> Plans that do not recognize and articulate risks make it extremely difficult to assign proper value to space science investments. (NRC, 1997, p. 12.)

Measuring risk is an important step in evaluating programs, especially if the risks associated with spaceflight are to be driven down. Later sections in this chapter suggest that acceptance of higher levels of risk may not be necessary and that improvements in system reliability are indeed possible in cost-constrained environments. Earlier RAND research has established that the hopes of realizing such a goal depend upon a high-level priority on risk reduction. (Alexander, 1988, p. 9.)

---

[2]Failures can have effects disproportionate to the size of of the mission, as illustrated by the loss of Argentina's Scientific Applications Satellite (SAC-B). The November 4, 1996 loss of SAC-B led to a diplomatic incident that was unexpected in the policy community. The next spacecraft in the series, SAC-C, was shifted from the Pegasus XL launch vehicle to the lower-risk Delta, where it remains comanifested with the New Millennium EO1 spacecraft.

PROCESS IMPROVEMENT CONSIDERATIONS

Process improvement happens behind the scenes, reordering and restructuring people and infrastructure in a constant drive for performance and cost-effectiveness. Sometimes, process improvement involves the modification of existing practices; other times, it means abandoning old methods in favor of new ones. While process improvement can result in upset, many private companies have learned to embrace it as an essential element of the quest for global competitiveness.

From the outside, process improvement can appear fickle. The PA field, for example, has adopted many new processes. In the early 80s, PA organizations were immersed in "quality circles," "statistical process control," and "total quality management." The 90s saw a transition to "benchmarking" and "quality action teams." Rather than signaling uncertainty in approach, this willingness to adopt new paradigms is a trademark of process improvement.

It is sometimes assumed that process improvement is relevant only for large commercial firms involved in high-volume production. The goal of process improvement, however, is to reduce cost and increase efficiency, regardless of how many items are produced. This was illustrated in a recent U.S. Navy survey conducted as part of the Joint Strike Fighter Program. The survey involved 17 major aerospace facilities to estimate the effect of changing design and development processes to emphasize producibility. The results of the survey estimated that a savings of approximately 25 percent could be achieved in the cost of the first production item by process improvement alone. (Smith, 1997, p. 2.)

A revolution has been under way for decades in the commercial electronics industry and, to a somewhat lesser extent, the automotive sector. Firms compete on the basis of performance and price but, more recently, also on the basis of quality and reliability. More than two decades ago, the Intel Corporation, facing a wide range of competitors and concerned about product reliability, embarked on an expansive process-improvement effort centered around the need for higher-quality products. Now the world's leading producer of advanced microprocessors, Intel produced some impressive results:

- A sixteenfold increase in microprocessor complexity (as measured by the number of transistors contained within each component) within five years

- Three orders-of-magnitude improvement in quality

- A fourfold increase in productivity (as measured in terms of number of units manufactured per employee).

There are many technical reasons for this success, but a critical factor was a decision to commit the corporation to "an exhaustive effort to achieve world class quality." (Intel, 1996, p. iv.)

Defining Process Improvement

A technical enterprise can be thought of as a nested set of processes interacting to design, develop, test, and field a product. Organizations committed to process improvement layer these processes with practices that, in an ongoing sense, evaluate end-item performance and recommend improvements. Enlightened organizations take the next step, embedding these oversight functions into the core of the production flow: The result is a highly integrated, highly "self-sensing" entity with a vision of growth. (Hammer, 1996, p. 80–83.)

The creation and uniform application of metrics is an integral part of a commitment to process improvement. (Hammer, 1996, p. 16.) In the private sector, firms must continuously reduce costs and increase performance to retain market advantage. Simultaneously, they must improve reliability, a parameter they carefully calculate and monitor. Cost, performance, and reliability form a triad for measuring the success of improvement initiatives. Failure to measure any one of these parameters clouds evaluations of process improvement.

Process Change in Relation to Small Spacecraft

NASA has already committed itself to continuous process improvement and the development of the required metrics. (NASA, 1996c, Section 3.2.7.) The many discussions and site visits that were conducted during the course of this study made it clear that process improvement has begun in spacecraft programs. The steps being taken, however, are early steps. Many organizations are just beginning to create the internal metrics needed to evaluate success, the results of which are beginning to be reported in the general literature.[3]

Change has come quickly to spacecraft development organizations. Rather than have time for fundamental process redesign, most small spacecraft efforts have had to streamline and reshape existing practices rapidly in an effort to cut cost and development time. Risk management is an example of a process for which sweeping changes have not yet settled out. NASA is now creating new processes that reflect the exigencies of tight budgets, but in the interim, small spacecraft programs have had to abandon tradition and create new strategies for approaching risk. The result is a wide variation in approaches to risk, in large part determined by the style and experience of the spacecraft team. (Gindorf et al., 1994a, p. 20.) The type of mission and the incentives under which the spacecraft team operates have also influenced the approach taken.

---

[3]Small spacecraft programs are closely evaluating the impact of new technical and management practices, including open discussion of costs. For an example of these self-evaluations, see Strope (1996) and Hemmings (1996).

In terms of the processes used to build them, today's small spacecraft are more similar to those of the past than will be likely in the future. What has been termed the "smallsat revolution" is in many ways a streamlining of traditional methods and an application of evolving technologies and techniques. For example, analog systems in spacecraft are only now being replaced with higher-performance digital designs, a transition long under way in terrestrial products. Part of the reason for this is cultural, part is sheer practicality (instruments, for example, remain predominantly analog), but the larger reason is that new approaches and methodologies are only now beginning to debut.

There are many signs that the way spacecraft are designed, built, and operated could be dramatically improved through new processes. A potential way to accelerate the adoption of new approaches might be for NASA to link its commitment to process improvement with a measurable vision of future spacecraft cost, performance, *and* reliability.

Measuring Process Improvement

Like industry, government shares an interest in reducing cost and increasing performance. At NASA, significant work has already been accomplished in measuring program performance, and spacecraft technical performance is relatively easy to measure. New full-cost accounting practices are also maturing rapidly to provide an even playing field for program evaluation. It is not clear, however, that reliability or risk reduction is a parallel objective within the government. If cost is reduced and performance increased, but risk is not at least held even, new approaches will be difficult to label as improvements.

THE ELEMENTS OF RISK

In the private sector, it is relatively easy to evaluate the success of a new approach. This is especially true when it comes to measuring reliability. Large production runs support the compilation of the statistical data needed to predict reliability accurately. For science spacecraft, which are one-of-a-kind items that use many components built in limited number, accurate prediction of reliability is a substantially greater challenge.

Risk management practices are currently receiving a great deal of attention and review within NASA, and the focus of these efforts is on new ways to measure risk and reliability quantitatively. The goal is to develop a set of quantitative tools that allow risk to be uniformly measured and predicted. This will be discussed below, following a brief review of definitions.

Some Risk Definitions

Risk is a higher-order term than reliability; a manager will describe program risks, while an engineer refers to component reliability. An overall assessment of risk requires an understanding of the reliability of the individual mission components (spacecraft, launch vehicle, ground system, etc.) and also the potential challenges associated with mission cost and schedule. Risk and reliability can be defined as follows:

- *Risk* is the likely variability of future returns from a given project. The total corporate risk is the sum of the political, economic, ecological, social, and technical risks to which the project is exposed. It requires an understanding of the elements that drive risk, the probability that they might occur, and the impact they could have. (Ayers, 1977, p. 24.)

- *Reliability* in terms of a mission, is the probability that at least the essential elements of a system will survive to meet scientific objectives. (Hecht, 1992, p. 704.)

The relationship between risk and reliability is complex. Certainly, the lower the level of a spacecraft's reliability, the higher the level of risk for the mission. Yet spacecraft reliability is only one component in an overall evaluation of risk. Flying a low-reliability spacecraft will increase the risk to the mission. Yet, alone, a high-reliability spacecraft cannot mitigate mission risk if the reliability of the launch system is low, or if the mission objective exposes the spacecraft to unknown environmental effects. An adequate treatment of risk in a space mission requires attention to all risk areas, including how risk is perceived in the broader context of public acceptance.

Reliability, in turn, is also determined by many factors. Very low-cost spacecraft can take a high-risk developmental approach and still prove to be reliable if they pursue simple designs and use fewer components. For example, out of 41 launches, only one AMSAT satellite has failed to operate correctly on-orbit. (Larson, 1996, p. 51.) Most small NASA spacecraft are appreciably more complex, however, and it is known that reliability and complexity are inversely proportional. (Hecht, 1996, p. 14.) Reliability is also highly dependent upon how a spacecraft is designed and tested. A robust design can accept repeated failures and still prove to be reliable, while fragility in even the most expensive design can lead to catastrophic loss.

Risk in Past Programs

Risk has traditionally been viewed as a consequence of any human endeavor, amplified in the case of the space program by the complexity and dangers associated with missions of exploration. In the past, a principal goal of a spacecraft program was *risk avoidance*, a practice that required a large percentage of the budget. The approach to avoiding risk was "rule based." As mentioned in Chapter Two, spacecraft were in the past classified according to the rules spelled out in NMI 8010. Spacecraft complexity, launch constraints, expected lifetime, and other technical parameters were used to select a classification. There was, of course, considerable discussion and mutual agreement on risk among scientists and program managers. But once selected, classification implied hard reliability targets. Both NASA and DoD management handbooks outlined the various technical approaches to avoiding risk.[4]

New Risk Management Approaches

NASA spacecraft programs are now shifting to a perspective that views *risk as a resource*. (Greenfield, 1997, p. 8.) Since NASA can no longer afford to avoid it, risk has evolved from a prescribed design point to a dependent variable included in the various trades that are a normal part of planning and executing a space mission. In addition to necessity, the formulation of a new approach to risk management was also a response to the desire to imbed considerations of risk more deeply in the project-management function and, most importantly, to have it occur very early in the mission-planning process. The treatment of risk as a resource is a change in strategy that focuses the spacecraft team on preparing an integrated plan to account for the many variables that affect a mission. This includes variables that are sometimes overlooked as spacecraft teams concentrate on solving technical challenges, such as managing international partnerships and responding to oversight requirements. Perhaps the most important aspect of treating risk as a resource is refinement of what is often an amorphous relationship between risk and the practical ways it must be traded to attain mission objectives. It is hoped that applying the techniques of risk management throughout the planning, development, and operation of a spacecraft will make this more of a science than ever before.

Treating risk as a resource also has some obvious downsides. Since risk avoidance is unaffordable, NASA's science missions, by implication, have a greater potential for failure. Although the loss of a smaller spacecraft is not as economically significant as, say, the loss of a large, billion-dollar mission, the impact to science may still be profound. Following the loss of large missions, like Mars

---

[4]Most relevant in terms of technical risk are Defense Systems Management College (1997), Defense Systems Management College (1990), and NASA HQ (1994), and NASA (1996a). The proposed European Program Management Standard ISO-14300 includes a final section on risk management practices in the concluding chapter. Many of these handbooks have been revised to reflect a movement away from risk-avoidance practices.

Observer and the recent Advanced Earth Observation Satellite (ADEOS), more and more scientists believe that smaller spacecraft offer a better chance of acquiring a steady stream of space data. (Kallender, 1997, p. 1.) This outcome is unlikely, however, if small spacecraft contain more risk than past missions. Indeed, the growing importance of small spacecraft would suggest that NASA give high priority to reducing risks below historical values. The effect of a mission loss on public and media interest in space must also be considered.

Choices made in the course of conducting missions can have far-reaching effects, underscoring the need for the adequate communication of risks. The anticipated costs and benefits of efforts to reduce risk, the effect of risk on the attainment of scientific objectives, and the influence of budget change on risk choices are examples of information that could be synthesized from NASA's risk as a resource activity and communicated to policymakers.

The evolution of risk from an assigned value to a variable that must be calculated implies that quantitative relationships can be devised to relate the many factors upon which an overall assessment of risk depends. Any potential use of risk as a metric in evaluating process improvement similarly depends on the successful creation of these quantitative relationships.

Measuring Risk and Reliability

In practice, the task of determining risk and reliability in space systems is very difficult. Statistically significant data are usually only available for parts and components, and this information often proves to be inaccurate. As a result, there is considerable disagreement within the spacecraft community regarding the ability to calculate overall spacecraft reliability with any degree of precision.

For spacecraft, calculating risk and reliability largely depends on the use of predictive techniques, such as the Taguchi Method and Bayesian Analysis. Some methods of prediction are based upon comparison with existing or past spacecraft designs. Others rely on limited or accelerated testing that is then extrapolated to yield some estimation of reliability. The challenge of accurately calculating reliability is all the more difficult in small spacecraft designs. Typically, there are few analogs available to formulate comparisons. Testing to establish reliability is expensive and time consuming and is usually beyond the means of small programs. Also, small spacecraft designs have often used commercial components, for which reliability information related to performance in space is less available.[5]

---

[5]The use of commercial parts in small spacecraft is discussed in Appendix D.

NASA realizes that new methods for calculating reliability and assessing risk are needed. Several programs are under way to improve the field of risk assessment and to refine new quantitative methods for establishing reliability with accuracy. To be useful in the broader context of process improvement metrics, however, these improved measures would have to enjoy widespread use among spacecraft development organizations. As mentioned previously, current methods of measuring reliability and assessing risk vary widely.

New quantitative risk-assessment methods are designed to be program tools. Time and effort will be required to translate these methods into a useful strategy for evaluating overall program performance. In the interim, however, it could prove advantageous to use actual spacecraft reliability data. Meaningful reliability targets are now used by small spacecraft missions to evaluate performance. The SAMPEX mission, for example, required a reliability of 99 percent in terms of the data delivered to the PI. Actual performance data indicated that the mission achieved 99.9 percent. (Watzin, 1996b, p. 10.) In general, spacecraft failures are less severe, and spacecraft are living well beyond initial design points. This is an achievement for which NASA does not now receive recognition.

The Cost of Reducing Risk

Data related explicitly to the cost of reducing risk are very difficult to obtain. Managers typically have a qualitative sense of how a particular initiative will improve the reliability of a product, but systems of measurement that generate quantitative data are rare. It is generally acknowledged, however, that the cost versus reliability curve is as shown in Figure 5.1. Space systems have historically operated in the upswept right tail of the curve. Earlier RAND research was unable to discern a relationship between cost and reliability in space systems. This was attributed to the very high reliability demanded, which obscured the effects of marginal investments, and to variations in cost practices among the various programs studied. The research did, however, establish that the cost of achieving high reliability typically approached 30 percent of the total cost of the mission. (Alexander, 1988, p. 92.) As a consequence of shifting to a strategy that views risk as a resource, future space systems will likely operate on a different portion of the curve.

There is evidence that even complex systems, such as spacecraft, can achieve reliability improvements through relatively small investments. Military systems offer a few comparative points. The F-18 aircraft, for example, achieved a doubling in reliability for a 12-percent increase in total cost. In some cases, reliability improvements came while program costs were being reduced. The

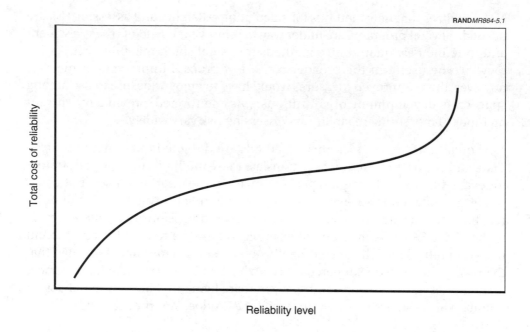

RAND*MR864-5.1*

Figure 5.1—The General Shape of the Cost-Reliability Curve

guidance system for the Minuteman missile demonstrated a 1500-percent reliability improvement while reducing the cost of procuring additional items by 50 percent. (Alexander, 1988, p. 37.) Early experience with spacecraft also shows that cost reduction and reliability improvement can be pursued simultaneously. The cost of the Air Force Agena-D was reduced by 30 to 40 percent while reliability improved by a factor of 6. (Katz, 1970, p. 16.)

The Concept of a Risk Portfolio

The Space Science and MTPE enterprises are diverse endeavors encompassing many discrete programs. Since NASA can no longer afford the risk-avoidance strategies of the past, it must now balance risk against cost and performance objectives. The level of risk will now be a returned value that will vary across the many missions that NASA funds. One notional way of viewing these endeavors, therefore, is as *enterprise portfolios* containing missions with varying scientific objectives, spacecraft sizes, producer organizations, *and* a distributed set of risks.

This model is analogous to an investment portfolio, which contains stocks, bonds, and funds selected to provide an optimal return on investment at some average level of risk. In a similar fashion, an enterprise portfolio would blend spacecraft program risks so as to maximize scientific return. The notional enterprise portfolio would also reflect the risks that the investor, the Federal government, is willing to take. A portfolio that, over time and for the same level of investment, produces the same level of scientific output is clearly desirable. Within the enterprise portfolio, cost, performance, and risk would form a symmetrical set of metrics useful in the preparation of the annual performance reports NASA must produce in response to the GPRA. An enterprise portfolio, however, could have several other advantages:

- *Enhanced communication.* Policymakers would have a direct mechanism for informing implementing organizations of their desires as investors of national resources. Risk can be directly related to goals and ultimately to the metrics used to evaluate success. The absence of such a linkage increases the subjectivity of developing and evaluating metrics. All levels of the enterprise have a common ground on which to discuss attainment of goals.

- *Better alignment of agency science and technology goals.* More-challenging scientific missions would carry a commensurably higher tolerance of risk. High-risk science missions would, by the nature of the objective, require technology beyond the state of the art, providing clear "technology pull" opportunities. This would assist research-oriented groups in building closer alliances with mission organizations.

- *Budget confidence.* Budgets can be aligned with more precision against the risk to be assumed by the implementing organizations. Higher-risk missions would be allocated more contingency funds, reflecting the uncertainty associated with integrating advanced technology, developing new spacecraft, and including possible backup options.

- *Identification of winners.* New technical approaches that result in high levels of reliability can be readily identified.

Within the enterprise portfolio, risk is a variable that becomes a metric of performance improvement. Creating an enterprise portfolio requires benchmarking current levels of risk across the spectrum of spacecraft missions. Once that is accomplished, however, the risk distribution within the portfolio becomes something that can be measured over time.

As a metric of enterprise performance, managed risk could become a useful device for focusing process improvement efforts. Figure 5.2 presents the notion that process improvement could be purposely applied to improving spacecraft reliability. Earlier, in Figure 2.4, a similar chart depicted the goal of increasing spacecraft performance through the use of advanced technology. In a similar vein, Figure 5.2 suggests that a goal of increasing spacecraft reliability could be achieved through purposeful process improvement.

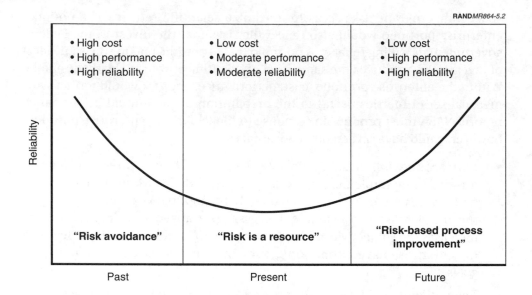

Figure 5.2—Goal of a Risk-Based Process Improvement Plan

Before the pursuit of such a goal could be considered, however, an important question must be addressed: Is it reasonable to expect additional improvements in the reliability of space systems? The following section examines four process-improvement trends that together could dramatically reduce the risk of space missions while permitting spacecraft to be designed with higher performance at lower cost.

KEY PROCESS-IMPROVEMENT TRENDS

This section explores four areas that could significantly change how spacecraft are designed, built, and deployed. This review will help address a fundamental question: Can processes be defined that allow spacecraft to be built at lower cost, with higher levels of performance, while also reducing the risk of failure? Four trends are outlined below that suggest an affirmative response. The four trends discussed below are covered in more detail in Appendixes B through E.

New Insights into Failures in Space Systems

Almost four decades of experience in building spacecraft and measuring the space environment have yielded a refined understanding of how to avoid failure. This experience is reflected in the fact that

•   The number of spacecraft failures has been steadily decreasing.

•   Failures, when they do occur, are less severe.

Within these trends, however, are some significant areas of concern that could affect continuous improvement in mission performance:

- Design-related failures are playing a more significant role as the total number of failures diminishes

- Mechanical failures contribute significantly to reduced performance or loss of spacecraft.

These areas deserve special attention in terms of focusing failure-reduction initiatives.

An area of great promise, in terms of understanding failure mechanisms and improving the construction of spacecraft, is the physics-of-failure approach. The goal of this approach is to replace empirical models of failure with more rigorous scientific analyses of how failure occurs in spacecraft components and subsystems. The increasing fidelity of failure models should aid in reducing the number of design errors and serious mechanical failures. The physics-of-failure approach is also very important in terms of (1) helping to predict the performance of new technology, (2) supporting the goal of overall risk reduction, and (3) increasing an awareness of reliability issues early in the design process.

To assist physics-of-failure initiatives, a greater degree of cooperation between the various organizations collecting and disseminating failure data is needed. The adoption of common recording and reporting formats would, for example, assist in the preparation of actuarial data. Funding for joint analysis efforts might be considered with the aim of providing a foundation for improved reliability and longevity estimates.

NASA has adopted a higher-risk approach in shifting to smaller spacecraft. One possible outcome is, in the short term, a higher rate of failure that disrupts current reliability improvement trends. Yet, as failure mechanisms are better understood, as the reliability of small launchers improves, and as small spacecraft programs incorporate high reliability systems, it is likely that future small spacecraft will continue the trend toward fewer failures.

Improved Test Processes

Despite its broad-scale importance, testing has long been an empirical process, with a great deal of variation in how the builders of space systems approach the test phase of a project. New insights into how components and systems fail have, however, illuminated deficiencies in traditional practices. The inherent quality and reliability of new components has also spurred a reexamination of test procedures.

Efforts to improve test processes were accelerated with then–Secretary of Defense Perry's 1994 decision to emphasize performance specifications in DoD procurement over long-standing military specifications and standards. Within the PA community, this decision was viewed with alarm in that milspecs form the backbone of traditional test procedures.

The arrival of smaller spacecraft also placed unique demands on test procedures: Testing had to remain effective yet be responsive to a smaller, tightly constrained budget; its criticality was increased, since many small spacecraft were proceeding with little or no redundancy; and a greater amount of new technology was being incorporated.

The response from the PA community has been a fresh look at existing test procedures aimed at replacing empiricism with experimental evaluations of the effectiveness of specific tests. Both NASA and the Air Force plan to establish refined test procedures that cost less to implement and are more effective in preventing defects from propagating. The net effect of these improvements is expected to be a significant improvement in the reliability of all space systems.

Development of High-Reliability Components and Subsystems

Space systems are clearly becoming more reliable. Improvements have been largely due to focused PA efforts within both NASA and the Air Force to bring developments in related fields into practice within the space program. Recently, the emphasis on high-reliability systems has increased. Manufacturers of spacecraft components are delivering products that demonstrate higher performance and greater reliability. In parallel, new design processes are incorporating reliability models that are more accurate and connected to failure-analysis databases. The commercial satellite communication sector, too, is placing greater emphasis on satellite reliability, in preparation for direct competition with terrestrial fiber networks.

Designing for reliability has a corollary effect—spacecraft tend to live well beyond original design points. Longer-lived spacecraft create a challenge in terms of operating budgets, yet longevity can be beneficial in terms of offering new approaches to conducting space research, observing the unexpected, having resources on hand to view emergent phenomena, supporting other missions, having greater mission-planning flexibility, and having more opportunities for training.

Of concern in terms of the future reliability of space systems is the increasing use of commercial components. High-quality commercial parts are more available than their space-rated alternatives and typically offer greater performance. At issue is the long-term reliability of electronic commercial devices encapsulated in plastic instead of traditional ceramics. Recent experience has shown that these components can meet the rigors of spaceflight, but certain areas, principally radiation resistance, remain that must be addressed through ongoing research. It is likely that supplies of space-rated components, especially high-performance microelectronics, will dwindle. Commercial electronic parts will, therefore, be increasingly important to the performance and reliability of future space systems.

Parts and components that are more reliable should translate into significant improvement in overall system reliability. Additionally, by their nature, small spacecraft offer advantages in terms of reliability. Smaller, more integrated systems have historically demonstrated higher reliability. The increasing proportion of microelectronic systems onboard future spacecraft should also lead to improved reliability, as should decreases in structural loads. It is possible, therefore, to envision future spacecraft that achieve unprecedented levels of performance through the use of systems designed expressly for high reliability.

Continuing Advances in Design Processes

The design of space systems is a comprehensive process that is being reengineered to deliver less-expensive, more-capable spacecraft that perform better and offer greater reliability. In regard to space systems, cost is the primary driver for changing the design process, since the design phase is typically the most expensive cost element in spacecraft TMC.

Builders of small spacecraft are especially pressed to minimize the length, and thus the cost, of the design phase. Some of the methods used to control design cost are:

- Capping the design effort (design-to-cost) and focusing on testing
- Forgoing the use of engineering test units
- Reducing new technology in the design.

These methods can work against other goals, such as reducing design-related failures and increasing the performance of spacecraft systems. New design approaches seek to improve the cost and technical effectiveness of the design process.

One of the most important improvements has been a greater degree of collaboration within design teams. The traditional hierarchical design process, built around the work breakdown structure, has been largely replaced by a collaborative process. RAND found that most of the small spacecraft programs in this study have reflected this shift by experimenting with or wholly adapting concurrent engineering practices and the use of integrated product teams.

Design process improvement has been paralleled by gains in the performance of modeling and simulation tools. Initial developments in this area have centered around the creation of design centers in which engineers are immersed in a team environment, surrounded by the latest computer-based tools. JPL's Project Design Center and Flight System Testbed are representative of such developments.

A natural extension of such centers is to connect geographically disperse teams via the Internet. Such "virtual" design environments connect teams via high-speed, fiber-optic links. Engineers can quickly analyze aspects of the emerging design by accessing local or remote tools, make changes, and communicate them to other team members.

The emergence of a collaborative design process, supported by computer-based environments containing advanced modeling and simulation tools, is an important development in terms of reducing the cost and risk associated with space systems.

NURTURING NEW METHODOLOGIES

The trends outlined in the previous section are naturally reinforcing. New insights into how systems fail lead to better design practices and the development of high-reliability components. The ability to design a spacecraft in a simulated environment reduces some of the test burden, and a focused test suite leads to a higher probability of catching defects and, therefore, a more reliable product. Substantial improvement in product performance and reliability can occur when trends reinforce in this manner. The confluence of trends is best ensured, however, when supported by purposeful planning and execution.

The previous discussion argued that reducing the risks associated with space missions is most readily accomplished when formally requested from above the program level. Metrics are also needed, as are achievable goals. A strategy to accelerate process improvement and reduction in mission risk might contain the following attributes:

- *Policy-level recognition.* To the extent that process improvement can reduce the cost, improve the performance, and reduce the risk of space missions, it reinforces national goals. High-level recognition would establish this awareness and help to energize organizations on the process-improvement track. Recognition might come in the form of an addition to the National Space Policy that set a goal for the improved performance and reliability of space missions. Line items in the budget that call out process-improvement initiatives in the budget might also help emphasize their importance.

- *Identification of metrics.* Individuals who create policy share a common cause with those who improve processes in regard to the creation of metrics. The policymaker needs meaningful measures to help evaluate the performance of the Federal enterprise. Process-change agents need metrics to evaluate the effects of new practices and techniques. Process-improvement initiatives could be viewed, therefore, as a source of useful cost, performance, and risk metrics.

- *Establishing achievable goals.* Metrics can be used to create a set of near- and long-term goals for process improvement. Goal setting would be a cooperative act designed to encourage the adoption of improved practices while ensuring that programs have sufficient time and resources to use them.

A strategy of process improvement would seek to bring new practices as rapidly as possible to small spacecraft programs, delivering more science at less risk in a constrained budget environment.

Expanding the Product-Assurance Role

Three of the four trends reviewed in the previous section—physics-of-failure analysis, new test practices, and high-reliability systems—are within the dominion of PA. Risk management, too, is essentially an assurance function. PA engineers are, therefore, well situated in terms of recommending alternative approaches and techniques. Notionally, NASA might consider expanding the role and mission of its PA function.

Figure 5.3 presents a means of envisioning an expanded PA role. Small spacecraft programs have made increased use of PA personnel through concurrent engineering practices. PA engineers now provide this *flight assurance* capability as an integral part of design teams and are involved very early in the process. This is a significant shift from the traditional oversight role of PA. More and more, however, this participation involves issues related to the use of new, and often unproved, technology. This *system reliability* function could translate into a new role for PA—bridging the gap between technology developer and the spacecraft engineer toward the goal of risk reduction.

Through an expanded involvement in the planning and review of NASA technology programs, PA engineers could help *assure* that maturing new technologies meet criteria for flight. The PA function could, for example, assist in refining TRL definitions and help create evaluation criteria for projects in the ITP. A higher level of PA integration might also help establish the "self-sensing" capability that serves to align process improvement efforts.

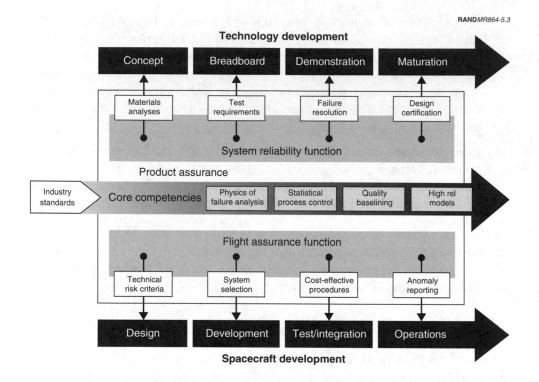

Figure 5.3—Potential Roles for Product Assurance

SUMMARY

NASA's space- and earth-science missions are important expressions of national policy, and the risks associated with these missions are a commensurately important matter. The shift to smaller missions has brought with it a change in risk-management strategy, from risk avoidance to managed risk. The implications of this change, most notably a higher overall level of risk, have not been communicated to the policymaker, mainly because a formal means of communicating risk desires and outcomes does not exist. The fact that the policymaker can no longer assume that risks are being minimized emphasizes the need for communication. Additional analytical work is needed to assess the effect of NASA's shift to a risk as a resource strategy. Future analyses might seek to

- Calculate the incremental cost of risk reduction strategies.

- Establish the cost of reliability in space systems (including an assessment of the most cost-effective means of achieving reliability), and the benefits that could accrue from such investments.

- Evaluate how changes in funding influence risk choices at the program level.

- Identify means of communicating and reporting the implications of risk choices.

Policymakers are also concerned with performance measurement. Evaluations are most valid, however, when they include measurements of performance, cost, and risk. In this regard, NASA's effort to refine risk measurement techniques has broader application. An improved ability to calculate risk supports efforts to measure performance, while also providing a mechanism for communicating risk.

It is, therefore, possible for policymakers to contemplate using risk reduction as a goal for the civil space program. Several long-term technical trends reinforce the notion that significant improvements in the reliability of space systems are possible.

Chapter Six

EVALUATING SMALL SPACECRAFT MISSIONS

Performance measurement is a pressing reality for all Federal programs. The 1993 GPRA amends Chapter 3 of Title 5 of the United State Code, requiring, for all Federal agencies:

- *A Strategic Plan*, to be transmitted to OMB by September 30, 1997 containing "general goals and objectives, including outcome-related goals and objectives, for the major functions and operations of the agency"; to be updated at least every 3 years.

- *Performance Plans* for each "program activity," containing performance goals, quantifiable objectives, resource requirements, and "performance indicators." (The agency can seek a waiver, under Section 1115(b), in areas where performance measurement is not feasible. Performance Plans are an annual submittal.)

Strategic planning within NASA's space- and earth-science enterprises is not a new experience. NASA's science programs are focused and peer reviewed, and the goals and objectives of its many elements are well articulated. For this reason, this chapter focuses on the challenges associated with preparing performance plans. The first section provides a review of comments from the RAND workshop related to metrics. The second section provides a notional mapping of performance metrics for measuring outputs of spacecraft missions. The chapter concludes with a discussion of the broader applications of peer and user reviews, particularly in regard to NASA's spacecraft technology program.

PERFORMANCE PLANNING AND METRICS FOR SMALL SPACECRAFT MISSIONS

If the strategic plan is seen as a destination, the performance plan might be viewed as the route plan and the record of the journey. It also provides a stable base for evaluation and for measuring the effect of process improvements. Elements of the performance plan must be traceable to both the OSS and MTPE

Enterprises' strategic plans, as well as the broader NASA strategic plan. This includes a reconciliation of resources (funding and people) against higher-order goals and objectives to ensure consistency in approach and balance among programs. Additionally, for NASA's response to GPRA to be meaningful, it is vital that there be a common understanding of the agency's goals among members of the administration and Congress. This includes an appreciation for risks associated with a cutting edge program like NASA's. (U.S. Senate, 1997.)

Workshop Results on Metrics

Metrics were discussed at length during the RAND workshop. Although participants unanimously understood and acknowledged the importance of performance measurement, most participants were reticent when asked to suggest specific measurement concepts. The participants urged caution in the application of metrics to spacecraft programs, recommending that metrics emerge from targeted discussions that include spacecraft developers, end users of scientific information, and members of the space-policy community charged with program oversight.

The cautionary stance of workshop participants reflects the complexity of defining and applying performance-based metrics to R&D activities. The GPRA legislation embodies simple concepts that are challenging to implement, especially when the object being measured involves the pursuit of science in space.

Performance-Based Metrics

The GPRA was designed to address a public perception that government programs are not working well. This perception may be at odds with the practical benefits of spacecraft programs (communication systems, intelligence gathering, weather prediction) and the profound impact that space- and earth-science missions have had on expanding human understanding of the universe. It may also be at odds with the cost reduction and performance improvements being demonstrated in small spacecraft programs. This is one perspective. Another is that the performance-oriented philosophy of the GPRA is an opportunity for both military and civilian spacecraft developers to show high returns on invested Federal dollars. The close tie between reporting information and being evaluated in terms of that information suggests that metrics should fully capture not only important scientific, technological, and educational outcomes but also the challenges of developing spacecraft programs.

NASA's strategic plan is a starting point for responding to the GPRA, but the development of the required performance plan, mapped to strategic goals and objectives, will be more problematic. Performance management has obvious challenges when applied to space-based R&D:

- The results, or outputs, of spacecraft-based research, although quantifiable in some sense, have only a tenuous relationship to eventual outcomes:
  - Knowledge evolves on a time line ranging from immediate to decades.
  - Research can lead to discoveries far beyond the hoped-for result.
  - Results combine in unexpected ways to yield new knowledge.
  - Outputs that appear equal are not always equal.
- Mission priorities must remain somewhat fluid to take advantage of emerging science opportunities.
- Space research is inextricably linked to a complex set of dependent performance variables (launch programs, operational components, etc.).
- The external environment is unstable.

The passage of the GPRA was done with the full awareness of the complexity of implementing it. To test its applicability, Congress established pilot programs to monitor agency responses. Only one of these pilot programs was an R&D agency: the Army Research Laboratory (ARL).

Satellite programs have components that are basic research, applied research, technology development, and production engineering, each with elements that are quantifiable and nonquantifiable. This suggests a hybrid performance plan in response to GPRA requirements. ARL created a "do what makes sense" performance plan that was a hybrid containing peer reviews, performance metrics, and evaluations of customer satisfaction. ARL sought 52 waivers from OMB under Section 1115(b) of the GPRA.[1]

A recent policy guide agrees with this assessment that NASA must create a hybrid performance plan, stating that "no single metric or group of metrics is likely to apply to NASA on a broad scale ... appropriate metrics have to be developed for different parts of the NASA research program." (NASA OSS, 1996b).

Table 6.1 presents a notional outline of what a hybrid plan might look like for spacecraft programs. The table suggests that performance depends upon success in three areas: research and analysis, flight programs, and technology

---

[1] Research Performance Measures Round Table (1995).

Table 6.1

Outline of a Possible NASA Hybrid Performance Plan

Program Element	Program Type	Section 1115(a)		Performance Criteria	Applied Metric	Peer Review	Review Period
		Apply	Waiver				
Research and Analysis							
Theoretical Science	Basic research		1	Enterprise strategic plan	Performance output	1	Annual
Mission Studies	Basic research		1	Enterprise strategic plan	Performance output	1	3 years
Flight Programs							
Mission Design	Development	1		Mission plan, PSR, PFP	Cost and schedule		Annual
Spacecraft Development	Development	1		Mission plan, PSR, PFP	Cost and schedule		
Mission Operations	Development	1		Mission requirements document	Measurement outcome		
Technology Programs							
Base Program	Applied research		1	RTOP	Survey	1	Annual
Focused Program	Development	1		STDB	Performance benchmarks	1	2 years
Flight Program	Development	1		Mission plan	Cost and schedule		Annual

programs. These areas contain basic research, applied research, and developmental components. Metrics can be developed for all of these areas, but in areas where discrete outcomes are difficult to establish, NASA can seek waivers. Some components of the plan can be traced to certain performance criteria. Table 6.1 identifies possible sources for the criteria used to develop metrics. Suggested metrics are also identified and explained in the following subsections. Note that peer and user reviews are suggested in all three areas, including technology programs.

The five applied metrics listed in Table 6.1 can be compiled in most cases from available data:

Performance Outputs. Although not the preferred performance measurement, output is often the only appropriate measurement for certain elements of R&D programs. NASA has used such measures as the "number of publications appearing in refereed publications for selected programs." (NASA Office of the Administrator, 1995.) Outputs based on counts are often contrived, offering little value for performance measurement. Count-based metrics also have the built-in tendency to focus attention on quantity instead of quality, often leading to efforts to "pump up the count." Count-based metrics can, however, be effectively paired with peer review.

Cost and Schedule. "On budget," "on schedule" performance is a reasonable expectation from space-science missions if adequate reserves and descope options are retained in the mission plan. Measures can be taken from NASA's Program Financial Plan (PFP), the Program Status Report (PSR), and detailed mission roll-ups. The ongoing development of a mission can be portrayed as an element of performance planning; the parameters might include the

- initial launch date
- scheduled launch date
- projected full cost
- cumulative spending
- percent descoped
- percent complete
- remaining reserve.

Measurement Outcomes. Measurement outcomes should be linked directly to mission performance objectives and requirements. For example, the principal objective of the MAP mission is full-sky mapping of the microwave background. A measurement outcome for this mission might be: "Complete a mapping of

100 degrees of sky at the wavelengths of 12 and 25 μm with a resolution of 0.21 mJy by 2001." Another example is listed in an OMB handbook for developing metrics: "Generation X observational satellite will successfully map 100 percent of the terrain of six Jovian moons to a resolution of 100 meters." (Groszyk, 1995.)

Surveys. Surveys offer an opportunity to get feedback for the program's "customer base." ARL used surveys effectively as part of an integrated approach to performance planning. Customers are asked to provide qualitative estimates of relevance, productivity, and quality. In the case of advanced space technology programs, the customer base is the end user—the spacecraft engineer who must implement new missions.

Performance Benchmarks. As Figure 6.1 illustrates, performance benchmarks represent the set of performance targets NASA's technology development program uses. Presenting these benchmarks requires a structure for organizing technology programs and agreed-upon targets that have been reviewed against technical and budgetary limitations. Such a framework exists in the form of OSS's evolving Space Technology Database. This database contains a work breakdown structure (taxonomy) for technology programs, along with performance targets for specific technologies. It also identifies linkages to flight programs when available. Performance benchmarks can be displayed graphically, offering an easy to understand performance tracking strategy.

THE NEED FOR REVIEW

The recommendations of the various NASA advisory committees and the growing importance of performance management suggest that expanded use of peer review makes sense. From the standpoint of efficient use of resources, and in consideration of the new requirements of the GPRA, peer and user reviews can be a powerful tool to solidify NASA's base of support.

Review is especially important in terms of advanced spacecraft technology programs. Although reviews of technology programs do occur within NASA, peer and user reviews have not been consistently applied and not with the rigor traditionally associated with science programs. In 1990, the Augustine Committee recommended that "NASA utilize an expert, outside review process, managed from headquarters, to assist in the allocation of technology funds." (Augustine Committee, 1990, p. 31.) The NASA Advisory Council's Federal Laboratory Review Task Force also recommended peer review of the agency's technology program, noting that external reviews should be applied to projects beyond the R&D phase. (NASA Advisory Council, 1995, p. 4.) RAND workshop participants were unanimous in their agreement with these recommendations.

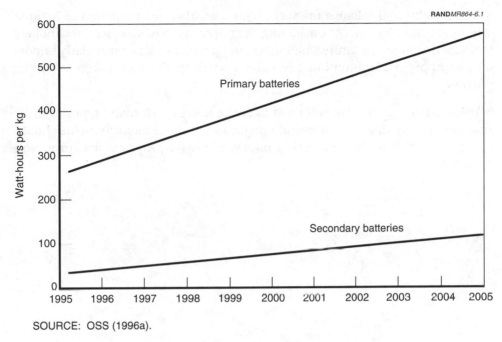

RAND*MR864-6.1*

SOURCE: OSS (1996a).

Figure 6.1—Battery Performance Benchmark

One of the principles of Cost of Quality management is that performance measurement should include the end user of the process. At NASA, this would mean placing the people who use space technology in an oversight position: mission scientists, who rely on technology to meet future requirements, and spacecraft designers, who must integrate new systems. The resulting process would contain both peer review and user review elements. The review process should also involve external, unbiased agents, who can assess the merits of the agency's technology transfer processes.

SUMMARY

NASA has taken significant steps toward improved accountability and performance measurement. Full-cost accounting practices will allow spending to be measured with precision. In terms of technical performance, NASA has an assortment of useful measures with which to build a performance plan. In particular and as described in Chapter Five, NASA is currently refining methods of measuring risk and reliability. Until these measures mature, current mission reliability data provide a useful way of measuring this important aspect of performance.

NASA's traditional reliance on peer review can also be broadened to include other program elements. Considering the importance of advanced technology to the performance of future small spacecraft programs, it is particularly important to apply a combination peer-user review to the technology planning process.

NASA's performance plan will most likely be a hybrid of many types of measurements. The final selection and integration of these measurements should strive to establish the connection policy, strategy, program structure, and resources.

Chapter Seven

CONCLUSIONS AND RECOMMENDATIONS

NASA's space- and earth-science programs are key elements of national space policy. They are also highly visible programs that are expressions of American scientific leadership and technological strength. Within this context, the renewed reliance on small spacecraft is an important development with far-reaching repercussions.

More than 35 years ago, small spacecraft took America's first steps into space. The size, complexity, and cost of robot explorers grew in proportion to the human hunger for knowledge, and small spacecraft were bypassed. Today, small spacecraft are returning, proliferating opportunities for research and helping to respond to a constrained Federal budget. The new generation of small spacecraft is proving that they can support a viable, although reduced, science program. There are also indications that, in the near future, small spacecraft could reach unprecedented levels of performance and reliability.

The results of this study and a RAND workshop on small spacecraft are summarized in the following conclusions and recommendations. The first section builds on the previous chapters to present study findings and conclusions, while the second contains recommendations that respond to the questions posed to RAND. Many of the findings of this study have implications for space missions of all sizes.

STUDY CONCLUSIONS

The introduction to this report presented a set of key questions policymakers have asked in regard to our increasing reliance on small spacecraft:

- What roles are small spacecraft currently playing in the civil space program?

- What strategies have proven especially effective in reducing cost and increasing performance of small spacecraft?

- What role does advanced technology play in the process of building small spacecraft?

- How should government evaluate civil small spacecraft programs to ensure that objectives are met cost effectively?

The answers to these questions will help policymakers make informed choices regarding the effectiveness of current efforts and the direction of future programs. Funding for space research will likely be constrained in the near term, so it is imperative that limited funds be expended as cost-effectively as possible. These questions were designed to build an understanding of the capabilities of small spacecraft and to shed light on how NASA uses them, now and in the future. NASA's space-science program remains a cornerstone of national space policy, and ways must be found to continue the pace of exploration. This underscores the importance of advanced technologies to enable more-capable missions within the available budget. Finally, measurements of performance are a new and vital part of all government programs.

To answer these questions, a study set of 12 missions was selected that represented the current class of small spacecraft missions in the astrophysics, earth science, and planetary exploration disciplines. These missions had an average dry mass of approximately 500 kg and have typically taken less than three years to develop. With an average cost of just under $150 million, these missions are a dramatic departure from the large developments of the past.

Small Spacecraft at NASA

The emergence of small spacecraft is often associated with Federal belt-tightening, but the science community was calling for change in how science missions were planned and executed a decade prior to substantive budget reduction. Pressure from the science community to reduce the time to develop new spacecraft and increase the number of research opportunities peaked in the mid-1980s. This drove NASA to consider smaller missions, a trend that was accelerated by reductions in the space-science budget that began in 1992. In FY96, NASA invested approximately $1 billion in small spacecraft, one-quarter of the total spending on space and earth science. Today, NASA's science programs are heavily populated by small spacecraft missions. Near-term budgets for space science are expected to remain flat, so small spacecraft will in all likelihood remain a critical program element.

Small spacecraft, like the Mars Pathfinder shown in Figure 6.1, are proving themselves to be powerful research instruments, sometimes making measurements equivalent in precision and resolution to much larger spacecraft.

Figure 7.1—The Mars Pathfinder's Sojourner on the Ares Vallis

They are ideally suited to focused explorations and as precursors to more expansive investigations. Future plans calling for cooperative networks of small spacecraft seem especially important in terms of both conducting exciting science and the opportunity to further reduce costs.

Small spacecraft are an effective response to dwindling budgets because they cost less to develop. They should not, however, be automatically viewed as cost-effective. They usually cost more per kilogram to build and launch than larger spacecraft and generally incur more developmental risk.

RAND workshop participants felt that there was danger in being overly aggressive in pursuing small spacecraft options. It is important that spacecraft be appropriately sized within available budgets. The NRC's Space Studies Board has made similar observations in relation to planetary missions in two separate reports. (NRC 1995, 1996) In this regard, it is important to note that the language in Section (3)(d) of the National Space Policy recommends that NASA focus attention on smaller, more capable spacecraft. In lieu of the above findings and recommendations, preferred language would seek to ensure that NASA pursues scientific objectives with the best balance of cost and risk.

Effectiveness of New Techniques

Fiscal constraint has driven a "design-to-cost" focus into all spacecraft missions. Keeping costs under control has, however, had implications beyond the design process: Most small spacecraft contracts are fixed-price or performance-based contracts and are tightly managed. Traditional management structures have been largely exchanged for highly integrated teams and "as needed" reviews. *This streamlining has saved approximately 20 percent in the total cost of missions.*

Spacecraft design teams are willing to cast far afield for new ideas and designs that might yield a better design. There are many conferences and workshops to communicate new techniques, which are well attended by NASA, DoD, and industry participants. Design teams have also made extensive use of the Internet to coordinate and plan, exchange information, inform the public, and transfer research data.

The maturity of the commercial space sector has become an important element in the drive to reduce cost. Spacecraft systems and entire spacecraft buses are evolving to become commodities, even to the point that NASA can consider procuring them as commercial items under new regulations. NASA is also turning the management of science programs over to PIs to a greater extent. The ability of the private sector to produce science spacecraft reliably and NASA's increased reliance on PI-mode management of missions give rise, however, to concerns that the agency might not be able to maintain its core competencies. More analysis and study are needed to evaluate NASA's changing role as a builder and manager of spacecraft, especially how the agency will attract and retain the appropriate balance of highly skilled scientists, engineers, and managers. Strategies for exploiting NASA R&D facilities and strengthening the agency's scientific and technical capabilities should also be examined.

There were also indications that NASA is, in many ways, relearning techniques and practices that it had used before migrating to large missions. The Air Force, on the other hand, has continued to build small spacecraft since the 1960s, using many of the design, test, and procurement practices NASA is now starting to use to build the current generation of spacecraft.

There are some side effects from current streamlining efforts, indicating that missions may have reached a "lean limit." The traditional high-stress periods in space missions have, in many cases, been replaced by a constant state of team anxiety in an all-out effort to stay within cost and schedule. Many corporate teams also have found it difficult to generate sufficient funds to capitalize the new equipment needed to improve performance on future missions.

Spacecraft programs have also adopted a new approach to risk management. In the past, spacecraft programs avoided risk; now, risk is accepted and managed within the funding envelope. In many cases, this change in strategy means that small missions are exposed to higher levels of risk. *There is no assurance, however, that program-level choices regarding risk are aligned with policy-level expectations.* Additional study of the implications of NASA's new risk strategy is needed. Establishing a means of communication risk desires and outcomes is also important.

Although there have been examples of completely revamped techniques, such as the creation of the first small satellite production line, spacecraft of the current generation are more the products of engineering practices that have been simply scaled back. This leaves open the potential for deeper changes in the approaches used to develop future science spacecraft. Technology will likely bring about ever-higher levels of performance, but process improvement will likely be the function that integrates new concepts in a way that leads to parallel reductions in cost. Four key trends were identified that support process improvement: new design techniques; improved test methods; a more scientific approach to, and understanding of, spacecraft failures; and high-reliability components and systems. These trends also suggest that future spacecraft could achieve significant reductions in risk.

Advanced Technology

Advanced technology will likely be a major factor in boosting the performance and improving the cost-effectiveness of future small spacecraft. The extent of the impact of technology depends on the nature of the mission, but all scientific disciplines will benefit from the many projects now under way. NASA currently invests approximately 2 percent of its annual budget on spacecraft technology, a modest investment that must be carefully managed to yield the highest possible return.

Small spacecraft, which can be developed on shorter timelines, offer a distinct advantage in being able to more closely approach the state of the art. This is an important attribute in that small spacecraft are pressed to meet ambitious scientific goals. Incorporation of advanced designs incurs cost and schedule risks, however, creating a dilemma for resource-constrained small programs. This underscores the need for close cooperation between organizations building spacecraft and those developing advanced systems.

Coordination of the many technology projects NASA funds has proven difficult in the past. In large part, this stems from the agency's dual mandate to conduct basic research and develop focused technological applications. There are two technology cultures within NASA: One has an R&D focus, and the other is more closely aligned with developing new designs that respond directly to science mission requirements. The latter culture usually provides incremental improvements in technology. *Future mission requirements point to a need for more advanced technology, which, in most cases, lies far afield from the end user—the instrument or spacecraft builder.* Efforts to integrate advanced technology must bridge this cultural gap.

The study noted strong administration support for NASA's spacecraft technology programs. The effectiveness of this investment depends critically on the successful implementation of a new ITP. To avoid past criticism of agency technology planning efforts, the ITP must have a clear mandate. The ITP will bridge technology programs occurring in different NASA enterprises, several flight programs, and various field centers. The authority of the ITP over the planning efforts of these groups must be well defined, and a clear determination must be made as to how the ITP will effect funding of specific projects. Whether the scope of the ITP extends beyond spacecraft bus technologies to the development of instruments and sensors, ground systems, design processes, and test facilities must also be determined. The connection between future scientific mission requirements and individual technology projects must also be apparent, as must the performance milestones for evaluating each project. NASA's technology program also bears some relation to military and commercial spacecraft technology efforts. *The ITP must establish a clear means of communicating NASA program content and intent to external organizations to minimize overlap and ensure that Federal technology investments are being used to greatest effect.* Finally, implicit in the creation of the ITP is a high degree of trust between those building spacecraft and those developing technology. To firmly establish this trust, an assurance is needed that the methods and terminology that technologists use to measure and describe the readiness of new designs are relevant to, and understood by, the spacecraft builder.

To leapfrog to higher levels of spacecraft performance, NASA has made a substantial investment in dedicated technology-demonstrator spacecraft. Science is a minor objective for these missions; they focus instead on testing a suite of advanced new technologies. While the goal of fully exercising a suite of technologies in an actual mission environment is appealing, these missions can be problematic. Cost and schedule risks can be high, and they can be expensive to implement. Many alternative means are available for retiring the risks associated with a new design or system; historically, failures due to new technology have not been a significant concern. *Dedicated technology missions should be employed only when alternative approaches are not viable or when cost-effectiveness can be clearly demonstrated.*

Measuring Performance

Measurement of performance is a requirement of the GPRA. Continuous process improvement initiatives also depend upon accurate performance metrics. NASA spacecraft programs have a variety of measurement techniques with which to prepare a response to the GPRA. In measuring performance, however, it is important to apply a symmetrical set of metrics; otherwise, progress cannot be assured. NASA can readily assess scientific and technical progress by applying a variety of hard measurements. These represent performance benchmarks, cost and schedule milestones, or the completion of planned measurements by a spacecraft instrument. Hard measures can be complemented by softer measures, such as peer reviews and surveys. Full-cost accounting methods, which are rapidly maturing, should supply ample measures of cost and schedule performance. Missing from the set of metrics, however, is an evaluation of program risk or reliability. NASA is currently working to refine methods for quantitatively measuring overall risk. Until these measures are matured, NASA should consider using currently available data related to mission reliability.

RAND workshop participants recommended that NASA conduct an annual review of spacecraft technology projects and the ITP, echoing the recommendations of the 1990 Augustine Committee. It was also suggested that this review be managed by an unbiased outside agent and that the end users of technology products, the mission scientists and spacecraft and instrument developers, be included.

RECOMMENDATIONS

Small spacecraft will remain an important element of the space program for the foreseeable future. The potential exists for a renaissance in knowledge of the earth and space, within the confines of a limited budget. It is in the national interest to ensure that small spacecraft mature rapidly to become reliable human proxies on voyages of exploration that will continue to engage the human spirit. This study recommends several actions to strengthen and expand these programs:

Civil Space Policy Objectives

- Establish a goal within the National Space Policy calling for NASA to pursue mission excellence in the design and development of science spacecraft. This goal would formally recognize the important role that the agency plays in improving the performance of space systems, which, in turn, strengthens our leadership in both the military and commercial space sectors.

- Conduct a review of NASA roles and missions in relation to a mature commercial space sector. This review should seek to identify NASA's unique strengths and capabilities in the areas of technology development and spacecraft design, development, and operations. It should also clearly identify the functions that must be retained and reinforced in regard to agency core competencies.

Improving Technology Planning and Implementation

- Firmly establish the ITP as NASA's focal point for the coordination of all instrument, spacecraft, and ground-system technology initiatives. Merge current spacecraft technology-development programs under the umbrella of the ITP. Within the ITP, create guidelines that establish a balance between basic research and nearer-term development projects.

- As a supplement to the ITP, prepare an annual report for instrument and spacecraft research and development projects. This report should include budgets (past, current, and projected spending), milestone schedules, and performance benchmarks.

- Initiate requirements for future technology flight-demonstrator missions from within the framework of NASA's ITP. The definition of these missions should emerge from a process that validates that flight in space is the only method of adequately retiring the risk of using a new technology. Additionally, this process should validate that technology demonstration is being pursued by the most cost-effective means.

- Evaluate the use of incentive awards to spacecraft development teams for advanced technologies that can be matured, documented, and prepared for transfer to other spacecraft developments and/or terrestrial applications.

- Examine the potential for integrating the PA function into the technology planning and implementation process. The goal of this examination would be to evaluate whether product assurance engineers can assist technologists to prepare their products in a form most readily integrated by the end user—the instrument or spacecraft designer.

- Forge more cooperative alliances within the spacecraft-development community. Consider broader application of the partnership model the New Millennium Program created between science teams and developers of advanced technologies.

Risk Measurement and Reduction

- Increase funding for efforts to mature quantitative measurement of risk and reliability. New risk-measurement techniques should be designed to support not only the technical management of missions but also the need for NASA program offices to communicate risks to the policymakers.

- Direct additional funds to research in high-reliability space systems and to the study of failure analysis, new test practices, and advanced design processes. Additionally, augment funding for test and evaluation of high-reliability mechanical systems for small spacecraft.

Measuring Performance

- Apply relative measurements of reliability within the earth- and space-science portfolios to monitor process improvement. Also, apply these measures (a) to communicate overall program risk between NASA and policy offices and (b) to distribute reserves within the programs.

- Formalize NASA's process improvement by benchmarking current spacecraft programs in terms of spacecraft cost, performance, and reliability and relate progress in terms of the change of these parameters within the earth- and space-science portfolios.

- Create a formal review process for the ITP. The review should involve senior technologists as peers. It should also include individuals who use space technology—mission scientists, who rely on technology to meet future requirements, and spacecraft designers, who must integrate new systems. The resulting user-peer review process should also involve external, unbiased agents, who can dispassionately assess the merits of the agency's progress on these programs. Peer review results should be reported annually in the ITP report supplement.

SPACECRAFT COST COMPARISONS

A set of NASA missions was selected for this study that represented various stages of development and a diverse set of design approaches. The mission set included spacecraft built by large and small companies and by NASA GSFC and JPL. With the exception of Clementine, the mission set contained only NASA spacecraft. Clementine, a DoD mission designed to test military technologies, was included because it retained strong scientific objectives and was clearly designed to test low-cost design and production techniques.

MISSION DATA SET

The missions reviewed during the course of this study represented a mixture of objectives. Per the guidelines of the study, the spacecraft in the mission set had an average dry mass of under 500 kg. The set of 12 missions (13 spacecraft) included earth observation, planetary missions, and observatory spacecraft:

- Discovery: NEAR and Mars Pathfinder

- Explorer: SMEX SWAS and TRACE, and MIDEX MAP

- New Millennium: DS1 and EO1

- SSTI: Lewis and Clark

- Surveyor: MGS, Mars '98 (Lander and Orbiter)

- Clementine.

Some of the missions could be clearly classified as scientific (NEAR, SWAS and TRACE, MIDEX MAP, MGS, and Mars Surveyor '98); others (the New Millennium Deep Space 1 and Earth Orbiter 1, and SSTI Lewis and Clark) were primarily technology-demonstrator missions. The Discovery Mars Pathfinder mission represents an equal balance of science and technology objectives.

Absent from the mission set was the NASA SMEX Program's SAMPEX mission. SAMPEX, the spacecraft that marked NASA's return to small spacecraft development, was excluded by design. NASA invested heavily in the procedures and designs used to build SAMPEX with the express goal of changing the methodologies that existed at the time. SAMPEX was designed to meet the requirements of future SMEX spacecraft and also the XTE spacecraft; the mission cost was, therefore, expected to be above the norm. (Bearden, 1996, p. 40.) More than three times the civil-servant hours were invested in SAMPEX, for example, than in subsequent SMEX spacecraft. (Watzin, 1995, p. 8.)

ACQUIRING AND PREPARING COST AND TECHNICAL DATA

To enable a comparative review of small spacecraft development trends, a common definition of costs was needed. For this study *Total Mission Cost* (TMC) was used

> *Total Mission Cost* is the accumulated cost of a mission from inception (the point at which a proposal has matured into a defined new start) to completion (the predicted end of scheduled operations and data analysis). It includes in-house government (civil servant and support contractor) personnel costs; the estimated value of GFE; and all costs associated with design, development, integration, test, launch, mission operations, and data review and archival.

Definitions of TMC vary, but the one used in this study agrees closely with NASA's.[1] This definition of TMC includes the cost of integrating new technology into a mission but identifies it as a separate cost element from the development of the spacecraft and instruments. TMC includes all mission-unique ground support equipment (GSE), including test and integration equipment and control center hardware and software.

The challenge in accurately defining TMC occurs at the beginning and end of a mission: When is a mission first considered a mission, and when does it definitely end? Variations at the extremes, however, are not usually large determinants of cost. The approach used in this study was to capture mission costs following the completion of mission design (the traditional Phase A part of a mission) and up to "launch plus one year" of operations. Costs were broken down by phase (design, development, test, launch, and operations), by year, and by spacecraft system (power, communications, command and data handling, etc.). Costs associated with PA, system engineering, GSE, and GFE were also acquired.

---

[1]One of the most comprehensive definitions of TMC can be found within the cost guideline for NASA's Discovery Program. There, TMC is defined "as those costs necessary to accomplish all phases of the mission from Phase A through Phase E, regardless of the source of funding." The definition of TMC used in the study is broader than what NASA calls *Life-Cycle Costs*; see OSS Glossary at **http://www.hq.nasa.gov/office/oss/**.

In the majority of cases, the small spacecraft in the mission set were built by contractors, or by JPL, a federally funded research and development center. Data associated with these programs captured the full costs of designing, developing, launching, and operating the spacecraft. Data acquired for the Clementine mission also represented the full cost of the mission, since the Naval Research Center operates under Defense Business Operating Funds (DBOF) rules. However, NASA built several missions in house with contractor support. Full-cost accounting practices for government in-house spacecraft programs had not reached a sufficient level of maturity to be used in this study. In these cases, the number of civil-servant labor years was identified for each mission. These hours were translated into costs using a labor rate of $132,000 per professional year. It was recognized that this procedure could not ensure that all spacecraft programs were on an equal cost footing. However, potential errors were not judged to be significant.

Cost data for the Clementine mission were provided directly by the Naval Research Laboratory. Data for NASA missions were formally requested from the Office of the Chief Financial Officer. The only exception to this procedure was data for the SSTI Lewis spacecraft. No data were provided for this mission, and information contained in this report should be considered a best estimate only.

The mission set included completed spacecraft and ones under development. Although this introduced some uncertainty into the data, it was not judged to be a major source of error since small missions are maintaining strict cost caps. All costs were prepared in FY96 dollars using NASA inflation indices.

SPACECRAFT TECHNICAL PARAMETERS

A review of sources of spacecraft technical specifications revealed a great deal of variation. To improve the accuracy of these data, it was decided to acquire information directly from the respective program offices. Technical data related to the various spacecraft were acquired using a survey form that requested the information listed in Table A.1.[2] These data were refreshed prior to publication of this final report. Every effort was made to ensure that information for spacecraft still under development was accurate, but final values are likely to change slightly. In each of the following tables, a dash indicates a parameter that could not appropriately be applied; "n/a" indicates that data were not available.

---

[2]The survey instrument was distributed by fax and e-mail directly to the various program offices. Some of the fields in Table A.1 were calculated from the data provided, such as instrument mass fraction and solar array efficiency.

Table A.1

Spacecraft Technical Specifications

Mission	Spacecraft	On-Orbit Design Life (yrs)	Target Apogee (km)	Inclination (degrees)	Contractors (No.)	Instr. Mass (kg)	Prop/Con Mass (kg)	Bus Dry Mass (kg)	Total S/C Dry Mass (kg)	Total S/C Wet Mass (kg)
Clementine		n/a	Lunar/Comet	Ecliptic	1	8	223.0	227.0	235.0	458.0
Discovery	NEAR	4.00	Asteroid @ 2.2 AU	12.0	1	56	325.0	424.0	480.0	805.0
	Mars Pathfinder	0.70	Mars	Ecliptic	1	25	80.0	810.0	835.0	915.0
Explorer	SMEX-SWAS	1.00	600	70.0	1	102	0.0	185.0	287.0	287.0
	SMEX-TRACE	1.00	650	98.0	1	58	0.0	152.0	210.0	210.0
	MIDEX-MAP	2.25	Lunar-assist	L2	1	215	51.0	338.0	553.0	604.0
New Millennium	Deep Space 1	2.00	Deep Space	Ecliptic	1	10	91.0	269.0	279.0	370.0
	Earth Observer 1	1.00	705	98.2	1	64	15.0	216.0	280.0	295.0
SSTI	Lewis	3.00	523	97.4	1	81	12.0	195.0	276.0	288.0
	Clark	3.00	476	97.3	1	119	12.0	155.0	274.0	286.0
Surveyor	Mars Global Surveyor	6.00	Mars	Ecliptic	1	76	388.2	597.7	673.7	1,061.9
	Mars Surveyor '98—Lander	0.90	Mars	Ecliptic	1	23	64.0	527.0	550.0	614.0
	Mars Surveyor '98—Orbiter	4.00	Mars	Ecliptic	1	46	276.0	313.0	359.0	635.0
Baseline	RADCAL	1.00	LEO	89.5	1	9	0.0	82.5	91.5	91.5

Table A.1 (continued)

Mission	Spacecraft	Spacecraft Volume (m3)	Instrument Mass Fraction (%)	Launch Vehicle	Upper Stage	Bus Pointing Accuracy (degrees)	Bus Pointing Knowledge (degrees)	Stabilization Type	Number of Thrusters	Fuel Type
Clementine		1.92	10.6	Titan II	Star-37	0.0500	0.030	3-axis	12	Hydrazine
Discovery	NEAR	7.90	11.7	Delta II 7925	Star-37	0.1000	0.003	3-axis	11	Hydrazine
	Mars Pathfinder	2.90	—	Delta II 7925	Star-48B	1.0000	n/a	Spin	8	Hydrazine
Explorer	SMEX-SWAS	1.30	35.5	Pegasus XL	n/a	0.0008	—	3-axis	n/a	n/a
	SMEX-TRACE	1.00	27.6	Pegasus XL	n/a	0.0060	—	3-axis	n/a	n/a
	MIDEX-MAP	10.10	38.9	Delta 7325	Star-48	0.0300	—	3-axis	6	Hydrazine
New Millennium	Deep Space 1	4.40	22.9	Delta 7326	Star-37	0.2000	n/a	3-axis	8	Hydrazine
	Earth Observer 1	3.20	29.3	Delta 7320	n/a	0.0090	n/a	3-axis	4	Hydrazine
SSTI	Lewis	1.40	43.4	LMLV-1	Star-48	—	0.004	3-axis	8	Hydrazine
	Clark			LMLV-1	n/a	2.0000	0.020	3-axis	2	Hydrazine
Surveyor	Mars Global Surveyor	2.60	11.3	Delta II 7925	PAM-D	0.5700	0.180	3-axis	12	Hydrazine
	Mars Surveyor '98—Lander	3.00	—	Delta 7425	Star-48	n/a	n/a	3-axis	8	Hydrazine
	Mars Surveyor '98—Orbiter	1.60	12.8	Delta 7425	Star-48	1.1000	n/a	3-axis	8	Hydrazine
Baseline	**RADCAL**	**0.2**	**9.8**	**Scout**	**n/a**	**10.0000**	**5.000**	**Grav. Grad.**	**n/a**	**n/a**

Table A.1 (continued)

Mission	Spacecraft	Thermal System Mass (kg)	Power System Mass (kg)	Solar Array Material	Solar Array Area (m2)	Solar Array Efficiency (W/m2)	Solar Array Mount	Beginning of Life Power (W)	Average Power (W)	End of Life Power (W)
Clementine	Clementine	n/a	32.0	GaAs/Ge	2.30	156.5	Deployed	360	n/a	n/a
Discovery	NEAR	11.0	64.4	GaAs/Ge	8.90	211.2	Deployed	1,880	1,600	1,500
	Mars Pathfinder	30.0	23.0	GaAs	4.00	62.5	Body Mount	250	190	—
Explorer	SMEX-SWAS	4.3	59.0	GaAs	2.50	240.0	Deployed	600	270	525
	SMEX-TRACE	3.5	32.0	GaAs	1.30	203.8	Deployed	265	90	200
	MIDEX-MAP	40.0	41.0	GaAs/Ge	3.10	193.5	Deployed	600	320	400
New Millennium	Deep Space 1	8.0	108.0	GaInp2/ GaAs/Ge	9.00	288.9	Deployed	2,600	n/a	n/a
	Earth Observer 1	7.0	36.0	GaAs	4.50	140.0	Deployed	630	250	600
SSTI	Lewis	n/a	n/a	GaAs/Ge	—	—	Deployed	370	n/a	n/a
	Clark	1.5	11.0	GaAs	3.60	111.1	Deployed	400	165	350
Surveyor	Mars Global Surveyor	17.5	73.6	GaAs + Si	12.00	56.4	Deployed	677	645	624
	Mars Surveyor '98—Lander	n/a	40.5	GaAs	3.70	181.9	Deployed	673	n/a	307
	Mars Surveyor '98—Orbiter	n/a	46.3	GeAs	7.40	202.7	Deployed	1,500	n/a	515
Baseline	**RADCAL**	**0.1**	**19.0**	**Si**	**0.81**	**55.6**	**Body Mount**	**45**	**25**	**18**

Table A.1 (continued)

Mission	Spacecraft	System Power Density (W/kg)	Battery Type	Downlink Data Rate (Kbps)	Comm. Band	Transmitter Power (W)	Central Processor (Mips)	Mass Memory (MB)	Harness Pinouts (No.)	Lines of Flight Software Code (K lines)
Clementine		11.3	NiH2	128	S-band	5.0	1.7	1,600	—	—
Discovery	NEAR	29.2	Super NiCd	9	X-band	5.0	2.0	2,000	5,000	—
	Mars Pathfinder	10.9	AgZn	11	X-band	13.0	20.0	128	—	4,100
Explorer	SMEX-SWAS	10.2	Super NiCd	1,800	S-band	5.0	2.0	88	2,500	56
	SMEX-TRACE	8.3	Super NiCd	2,250	S-band	5.0	2.0	300	2,500	75
	MIDEX-MAP	14.6	NiH2	666	S-band	5.0	6.0	256	2,380	650
New Millennium	Deep Space 1	24.1	NiH2	10	X-band	12.5	20.0	128	5,000	5,000
	Earth Observer 1	17.5	Super NiCd	105,000	X-band	5.0	25.0	1,200	—	—
SSTI	Lewis	n/a	NiH2	n/a	S-band	n/a	8.0	2,000	—	—
	Clark	36.4	NiH2	2,500	X-band	18.0	8.0	2,000	—	23
Surveyor	Mars Global Surveyor	9.2	NiH2	85	X/Ka-band	25.0	0.3	750	12,256	18
	Mars Surveyor '98—Lander	16.6	NiH2	2	X-band	15.0	20.0	128	3,300	30
	Mars Surveyor '98—Orbiter	32.4	NiH2	111	X-band	15.0	20.0	128	2,500	30
Baseline	**RADCAL**	**2.4**	**NiCd**	**19**	**C-band**	**10.0**	**0.2**	**4**	**1500**	**8K**

AN AVERAGE SMALL SCIENCE SPACECRAFT

The cost data gathered for this study provide a portrait of an average small science spacecraft. This information could prove useful in evaluating the effects of future cost-reduction strategies and for comparison with later studies. Based on the data provided for this study, Table A.2 provides the average costs for each mission element, along with the data range.

A breakdown of spacecraft element cost by percentage of TMC is shown in Figure A.1. The average TMC of small spacecraft in this mission set was $145M. The spacecraft (bus, instrument, integration, and associated ground equipment) represents 60 percent of the TMC.

ACCOUNTING FOR COMPLEXITY

On a cost-per-kilogram basis, most small spacecraft are relatively more expensive than larger spacecraft. This was shown in Figure 2.5, where the cost of

Table A.2

An Average Small Scientific Spacecraft

Cost Element	Mean	STD DEV	Range
Management	6,638	5,808	20,185
Planning and Analysis—Phase A	4,820	8,195	25,837
Carrier/Bus	59,129	28,671	110,166
Structures & Mechanisms	9,623	16,078	57,734
Thermal	696	824	2,442
Attitude Control System (ACS)	10,235	7,265	23,448
Command & Data Handling (C&DH)	6,193	3,402	12,048
Electrical Power System (EPS)	7,225	6,574	25,272
RF Communications	5,549	4,095	12,745
Propulsion	5,705	3,858	11,268
Flight Software	2,726	1,692	4,338
Harness	771	531	1,671
Ground Support Equipment (GSE)	1,532	1,485	4,688
Government Furnished Equipment (GFE)	1,402	1,302	2,255
Other	7,473	4,911	13,493
Spacecraft Systems Engineering	2,271	2,187	7,543
Product Assurance	1,650	1,224	4,054
Parts Procurement	3,414	3,066	6,668
Contamination Control	138	160	300
Spacecraft Integration & Test	4,630	2,492	9,012
Instrument	20,826	12,499	45,200
Launch	31,379	14,434	36,637
Vehicle	30,498	14,422	35,600
Flight Integration & Checkout	881	879	2,760
Operations	11,556	13,851	46,280
Other	6,153	8,680	23,101
Ground Systems	3,737	1,840	5,488
Science	2,417	2,588	5,200

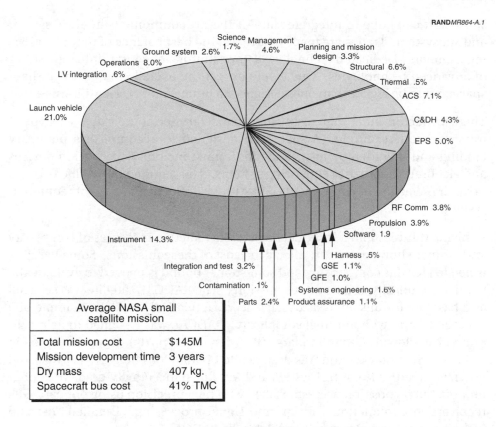

RAND*MR864-A.1*

Science Management
Ground system 2.6% 1.7% 4.6% Planning and mission
design 3.3%
Operations 8.0%
LV integration .6%
Structural 6.6%
Thermal .5%
ACS 7.1%
Launch vehicle
21.0%
C&DH 4.3%
EPS 5.0%
RF Comm 3.8%
Propulsion 3.9%
Software 1.9
Instrument 14.3%
Harness .5%
Integration and test 3.2%   GSE 1.1%
GFE 1.0%
Contamination .1%   Systems engineering 1.6%
Parts 2.4% Product assurance 1.1%

Average NASA small satellite mission	
Total mission cost	$145M
Mission development time	3 years
Dry mass	407 kg.
Spacecraft bus cost	41% TMC

Figure A.1—Average NASA Small Spacecraft Mission

missions in this study was plotted alongside an assortment of other missions. The fact that many small spacecraft cost more per kilogram than larger ones is not an unexpected result. Rather, as the mass of a spacecraft diminishes, it is reasonable to expect that development costs will remain somewhat elevated, reflecting a need to meet the continuing requirements of science. Also shown in Figure 2.5, however, are several small spacecraft that retain the cost-per-kilogram ratio of larger spacecraft. Complexity is the major reason for the cost variation in smaller missions.

Parameters Influencing Complexity

The complexity associated with a spacecraft is determined by many factors. As previously mentioned, small spacecraft programs are attempting to demonstrate ever-higher levels of technical performance. This usually requires that they integrate new, high-performance technologies, which is difficult to achieve in a cost-constrained environment. Typically, these missions are employing many state-of-the-art designs and components. To remain within cost caps, no spacecraft can hope to integrate state-of-the-art components in every system and subsystem. Trades are made to arrive at the best balance of new and existing elements. Dedicated technology demonstrators, of course, integrate suites of advanced technologies. One would expect the overall complexity of these spacecraft to be higher than ones designed to pursue purely scientific missions.

There are factors other than technology that determine complexity. The type of mission is one example. A spacecraft designed to accomplish a planetary landing and then deploy instrumentation must include deceleration devices and elaborate automatic sequencing systems. The required design life will also drive the selection of components and strongly influence the use of redundant systems.

Applying this reasoning, it should be possible to define measures of complexity and evaluate how complexity affects the cost of these missions. Some baseline is needed for this comparison, and some upper bound is needed to evaluate the relative complexity of the various missions. The RADCAL satellite was selected as a baseline for this analysis because it was a relatively modern example of a spacecraft built with minimal complexity. RADCAL was designed to help calibrate 77 worldwide C-band radars. (Bearden, 1996, p. 26.) It was launched in 1993 on a Scout rocket and was designed to operate for one year. RADCAL did not have an attitude control system and was flown in a gravity-gradient configuration, but its position was established with high precision using onboard GPS receivers in conjunction with ground signal processing. Detailed cost and technical data were also readily available for RADCAL.

A fictitious mission was used to establish an upper bound for the complexity of small spacecraft. Here, a small spacecraft was envisioned that combined state-of-the-art equipment in all systems and subsystems. When compared to the RADCAL baseline and this idealized example, each spacecraft in this mission set would demonstrate some relative complexity.

To evaluate mission complexity, a set of 11 parameters was selected. These parameters cannot be appropriately applied to all spacecraft in the mission set; exceptions are noted in the descriptions below. In other cases, complexity could not be calculated because spacecraft data were not available. While most of the parameters listed below supported direct calculation, judgment was required in some cases. The following measures were used in this analysis:

- *Design Life.* The required operating life of a spacecraft will drive the complexity of systems and the reliability of components selected in the design. Although not all missions require a long life, spacecraft designed to survive in space longer will usually be more complex. Six years was considered to be the state of the art.

- *Target.* The spacecraft in this mission set represent an assortment of destinations. The final destination of the spacecraft was selected as a measure, since it directly affects the complexity of the design. In this analysis, planetary landings were considered to be the most complex type of mission and earth-orbiting missions the least.

- *Spacecraft Density.* The density of a spacecraft can be considered a measure of complexity, since a greater degree of design is required to package systems and components tightly. A denser spacecraft requires a higher level of component and subsystem integration and careful attention to thermal loads.

- *Instrument Mass Fraction.* Increasing the instrument mass fraction (the ratio of instrument-to-spacecraft dry mass) is an expressed goal of many programs. This parameter is not applied to planetary lander missions, since these spacecraft pay a mass penalty in the equipment necessary to make a successful landing. It was also not applied to the NMP DS1 mission, since one of the goals of this mission is to demonstrate an ultralightweight instrument package for future microspacecraft. An instrument mass fraction of 50 percent was considered to be the state of the art for a small spacecraft.

- *Bus Pointing Accuracy.* The complexity of a small spacecraft is strongly influenced by the requirements of pointing accuracy. Some small spacecraft achieve extremely fine pointing accuracy, and the systems needed to provide this performance require a great deal of design and analysis. A pointing accuracy of 2 arc seconds (0.0005 degrees) was considered to be the state of the art.

- *Solar Array Efficiency.* Small spacecraft are using advanced solar array materials and production techniques. Solar array efficiency is a measure of the power output from the unit (watts) per square meter of area. To calculate solar array efficiency, the beginning-of-life (BOL) power output for each spacecraft was used. The state of the art for solar cell efficiency was determined to be 20 percent. When multiplied by the solar constant of 1,380 W/m^3, a upper bound of 274 W/m^3 was established.

- *Power System Efficiency.* Power-handling equipment for spacecraft now requires fewer components. Use of advanced designs is allowing spacecraft to deliver more power for a given mass of power-handling and conditioning equipment. Here, the state of the art was estimated to be 36 W/kg.

- *Downlink Data Rate.* Data-rate requirements for small spacecraft are strongly influenced by the type of mission and the instrument being flown. Some small spacecraft are pressing the state of the art in the use of high-data-rate communication systems. An upper bound of 10 Mbps was used in this analysis.

- *Central Processing Power.* In onboard processing, small spacecraft are leveraging the availability of commercial components and greatly expanding bus performance. Many spacecraft are using distributed designs with several high-speed processors. Although the processor itself is not a complex feature, the use of state-of-the-art processors usually indicates advanced data-handling designs. In this analysis, a state-of-the-art processing power of 22 MIPS was used.

- *Mass Memory.* Design complexity is also reflected in the amount of data the spacecraft must store and manipulate. An upper bound of 2,000 MB of mass storage was considered to be the state of the art.

- *Software Code Lines.* There is a wide variation in the amount of software used onboard a small spacecraft. Software is a important risk area, and some spacecraft are designed to minimize the amount of onboard execution code. Other missions place a great deal of reliance on onboard autonomy and will have elaborate software architectures. Future missions will likely depend more heavily on software for operation and health monitoring, and millions of lines of software code will be the norm. The state of the art for this study was set at 2 million lines of code, recognizing that many missions use far less.

Since many of the spacecraft studied are technology demonstrators, it is not surprising that, in some cases, spacecraft in the mission set were establishing the state of the art. It is recognized that some of the above measures bear a greater influence on the complexity, and therefore the cost, of a spacecraft. This was meant to be a first-order estimate, however, and the various measures of complexity were not weighted.

Calculating a Factor of Complexity

Table A.3 shows the final complexity calculations for each of the defined measures. A scale of 1 to 5 was used in this analysis to indicate movement from least to most complex. This scale approximates the range between the cost per kilogram of RADCAL and the average for missions in the data set. From these data, a final Factor of Complexity, F_C, was calculated for each mission as the unweighted average of the computed complexities.

Table A-3

Factor of Complexity Calculations

Mission	Spacecraft	On-Orbit Design Life (yrs)	Target Apogee	Spacecraft Density (kg/m³)	Instrument Mass Fraction (%)	Bus Pointing Accuracy (degrees)	Solar Array Efficiency (W/m²)	Power System Efficiency (W/kg)
Clementine		n/a	4.9	1.6	1.1	4.6	2.8	1.9
Discovery	NEAR	3.5	4.6	1.0	1.2	3.0	3.8	4.1
	Mars Pathfinder	1.0	5.0	5.0	—	—	—	1.8
Explorer	SMEX-SWAS	1.2	3.0	3.6	3.6	5.0	4.3	1.8
	SMEX-TRACE	1.2	3.2	3.4	2.8	4.9	3.7	1.5
	MIDEX-MAP	2.2	3.8	1.8	3.9	4.7	3.5	2.3
New Millennium	Deep Space 1	2.0	4.9	n/a	—	2.9	5.2	3.5
	Earth Observer 1	1.2	1.0	1.0	2.3	4.8	2.5	2.7
SSTI	Lewis	2.7	1.0	2.9	2.9	n/a	n/a	n/a
	Clark	2.7	1.0	3.8	4.3	1.5	2.0	5.0
Surveyor	Mars Global Surveyor	5.0	4.3	4.9	1.1	2.7	1.0	1.6
	Mars Surveyor '98—Lander	1.2	5.0	2.9	—	—	3.3	2.6
	Mars Surveyor '98—Orbiter	3.5	4.3	3.7	1.3	2.0	3.7	4.5
Baseline	RADCAL	1.0	LEO	92.0	10.0	10.0	56.0	4.0
Upper Bound	State-of-the-Art	6.6	Landing	288.0	50.0	0.0005	274.0	36.0

Table A.3 (continued)

Mission	Spacecraft	Downlink Data Rate (kbps)	Central Processor (Mips)	Mass Memory (MB)	Flight Software Code (K lines)	Complex. Counts	F_c
Clementine		1.2	n/a	4.2	n/a	8	2.8
Discovery	NEAR	1.0	1.4	5.0	n/a	10	2.9
	Mars Pathfinder	1.3	4.8	1.4	5.0	8	3.2
Explorer	SMEX-SWAS	3.4	1.3	1.2	3.1	11	2.9
	SMEX-TRACE	4.0	1.3	1.6	3.6	11	2.8
	MIDEX-MAP	1.9	1.9	1.5	4.5	11	2.9
New Millennium	Deep Space 1	1.0	4.8	1.2	5.0	9	3.4
	Earth Observer 1	5.0	5.0	4.6	n/a	10	3.0
SSTI	Lewis	n/a	2.3	5.0	n/a	6	2.8
	Clark	4.3	2.3	5.0	1.6	11	3.0
Surveyor	Mars Global Surveyor	1.1	1.0	2.5	1.4	11	2.4
	Mars Surveyor '98 —Lander	1.0	4.8	1.2	1.6	9	2.5
	Mars Surveyor '98 —Orbiter	.1	4.8	1.2	1.6	11	2.9
Baseline	**RADCAL**	**19**	**0.2**	**4.0**	**8.0**		
Upper Bound	**State-of-the-Art**	**10,000**	**22.0**	**2,000**	**2,000**		

Normalizing Spacecraft Cost per Kilogram

Table A.4 shows the spacecraft cost for each mission. "Spacecraft cost" includes design and development of the bus and the instrument, integration, management, and all other costs associated with delivering the spacecraft for launch. The "Cost per Kilogram" is simply the spacecraft cost divided by the dry mass. The "Normalized" values are the cost-per-kilogram values divided by the computed F_C.

A Complexity Cost Estimating Relationship

The computer value of F_C can be used to form a simple cost estimating relationship (CER). A CER based on dry mass, M (kg) and F_c is shown in Figure A.2. The cost of small spacecraft correlates well with dry mass, when corrected for complexity.

Table A.4

Normalized Spacecraft Costs per Kilogram

Mission	Spacecraft	F_c	S/C Cost (FY96 $M)	S/C Dry Mass (kg)	Cost per kg ($M/kg)	Normalized Cost per kg ($M/kg)
Clementine		2.8	61.9	235	0.263	0.095
Discovery	NEAR	2.9	95.5	480	0.199	0.070
	Mars Pathfinder	3.2	165.3	835	0.190	0.063
Explorer	SMEX-SWAS	2.9	78.4	287	0.273	0.095
	SMEX-TRACE	2.8	55.0	210	0.262	0.092
	MIDEX-MAP	2.9	107.1	553	0.194	0.067
New Millennium	Deep Space 1	3.4	81.9	279	0.294	0.086
	Earth Observer 1	3.0	70.4	280	0.252	0.084
SSTI	Lewis	2.8	39.0	276	0.141	0.050
	Clark	3.0	37.6	274	0.137	0.045
Surveyor	Mars Global Surveyor	2.4	109.3	673.7	0.162	0.067
	Mars Surveyor '98—Lander	2.5	77.1	550	0.140	0.056
	Mars Surveyor '98—Orbiter	2.9	77.1	359	0.215	0.074
RADCAL		1.0	4.6	91.5	0.051	0.051

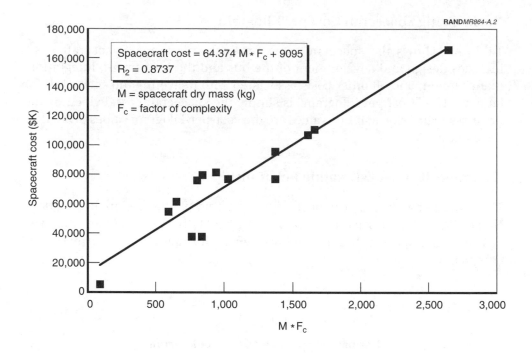

Figure A.2—CER Based on Complexity

SUMMARY

The normalized cost per kilogram numbers from Table A.4 were plotted in Figure 2.6. The distribution of F_C among the missions generally followed expectations. The complexity of technology-demonstrator missions was, for example, somewhat higher than that of science missions. Planetary missions, however, were considered to be underrepresented. These missions typically contain elements that are difficult to quantify. Mars Pathfinder, for example, is a combination of three elements, one of which is a sophisticated microrover. The calculations here were not designed to reveal this type of complexity.

When applied to the original Cost per Kilogram data, F_C produced a uniform distribution with a lower variance than the original cost per kilogram values. Relative to one another, however, there remains a significant distribution among these missions. This could be caused by an incomplete set of complexity measures or by variables unrelated to the complexity of the respective spacecraft.

Comparison to a different baseline mission could significantly change the computations of F_C. Additional studies might seek to broaden this analysis using other baseline missions with actual costs around $50,000 to $75,000 per kilogram.

FAILURE IN SPACECRAFT SYSTEMS

INTRODUCTION

Since the early days of the space program, spacecraft reliability has been steadily improving. Failures, when they do occur, also tend to be less significant.[1] There are, of course, some significant exceptions to the trend. Expensive spacecraft have been lost or impaired by single events that escaped detection prior to launch. Yet the overall trend is toward spacecraft that are more reliable and resilient. In large part, this trend is due to improvements in spacecraft components and subsystems and to the fact that the space environment has been characterized with greater accuracy.

In the future, spacecraft failures are expected to continue to decline. This appendix will provide an overview of spacecraft failure causes, examine some important failure trends, and assess the potential impact of a new tool, the *physics-of-failure* approach, in terms of helping to bring about further improvements in the reliability of space systems.

FAILURE IN SPACECRAFT SYSTEMS

Failures that lead to system anomalies and breakdowns are to be expected in any electromechanical system. For terrestrial systems, engineers can often test devices to the point of failure to evaluate a design; when failures occur in service, components can be recovered and studied. In contrast, expensive spacecraft systems are rarely tested to failure. To analyze failures during a mission, engineers must rely on telemetry, ground-test data, and operational analysis. Only rarely can components be recovered. Failure analyses are complemented

---

[1]A failure is generally considered substantial or significant if it causes a loss of 33 percent or more of the spacecraft's mission objectives.

by measurements of the space environment that aid engineers in the association of cause and effect.

To support the general study of spacecraft failures, data for individual missions are usually compiled in a database. Currently, there are four main repositories of spacecraft failure data: JPL's Payload Flight Anomaly Database (PFAD), NASA GSFC's Spacecraft Orbital Anomaly Report (SOAR), and the Air Force's Orbital Data Acquisition Program (ODAP) and the Space Systems Engineering Database (SSED).[2]

These data repositories provide a valuable historical record, but the task of acquiring and analyzing failure data is complicated by the lack of common reporting schemes and techniques. Procedures vary between NASA field centers, the Air Force, and industry. Even offices within a given organization can apply different techniques. Various organizations also differ in the way they treat failure events. Some choose to capture every event, no matter how minor, while others maintain some reporting threshold. In some reporting systems, a single failure event can have multiple assigned causes. The number of statistical data points may, therefore, exceed the actual number of reported failures. Most often, though, the cause of failure can be attributed to a single failure category.

The lack of commonality in the bookkeeping of failure data hampers the ability of PA engineers to monitor trends and focus research of spacecraft failure mechanisms. Resource-limited small programs have reported finding it difficult to sort through various failure archives to locate relevant information and apply lessons learned from previous missions.

Classifying Failure

Despite various approaches to classifying failure data, it is possible to create broad categories of failure and to review how manifestations of failure have changed over time. For this purpose, failures can be classified as (1) events caused by the space environment, such as radiation damage to circuits; (2) incidents for which some aspect of the design was inadequate; (3) problems with the quality of the spacecraft or of parts used in the design; or (4) a predetermined set of "other" failures, which include operational errors. A significant number of incidents cannot be attributed and are simply classified as "unknown."

It should be noted that failures are not always unexpected events. Mission timelines and cost factors sometimes demand that a spacecraft, such as

---

[2]The ODAP and SSED databases are maintained for the Air Force by the Aerospace Corporation.

Voyager, be launched with known problems. Engineers rely on robust designs, redundant systems, and prescribed workarounds to deal with anomalies that are deemed likely because they occurred, and were characterized, during ground testing. Depending on the reporting schema, expected problems may or may not be classifieds as failures.

Failures Caused by the Space Environment

The space environment provides an assortment of hazards whose ill effects can range from degraded performance up to catastrophic loss of a spacecraft. Some hazards involve impact destruction of spacecraft components, a particular problem in LEO.[3] Meteoroids, consisting mainly of comet remnants, and orbital debris fall into this category. Orbital debris consists of spent rocket components, launch and deployment fragments, and inactive payloads. The impact of particles weighing less than a gram can severely damage systems; heavier objects can completely destroy a spacecraft. The Russian Kosmos-1275 spacecraft was believed to have been destroyed in 1981 by a direct hit from a large piece of orbiting debris. Recently, the French Cerise satellite was crippled after being hit by debris (David, 1997a, p. 2), while an avoidance maneuver steered the European Radar Satellite (ERS-1) from a collision with the Russiam Cosmos 614 satellite (Selding, 1997, p. 1). Usually, however, very small particles are involved, leading to erosion and degradation of materials. Space Shuttle Orbiter windows are routinely replaced because of orbital debris damage, and erosion and penetration degrade solar-panel performance over time.

Although orbital debris poses a serious threat to manned and unmanned spacecraft, it remains less statistically significant than other environment factors. For LEO spacecraft, the tenuous upper atmosphere generates significant asymmetric drag on a spacecraft, a force that varies with the solar cycle, which must be countered by onboard propulsion systems. Atomic oxygen in the upper atmosphere can also cause serious deterioration of spacecraft materials and coatings.

Variations in solar and albedo radiation lead to a dynamic spacecraft thermal environment. Thermal and radio frequency interference effects are increasingly important in small spacecraft, since smaller volumes increase the sensitivity and susceptibility of parts and equipment to radiated emissions. Smaller spacecraft also tend to operate at higher computational loads than earlier

---

[3]Potential impact damage is a significant concern for low-flying assets, such as the Space Station, and for the LEO communication satellites now being deployed in great number. A recent study concluded that there is a 50-percent chance of a collision within 5 years for large spacecraft constellations; see Glicksman (1996), p. 6.

spacecraft. As a result, they can run hotter and drive temperatures at thermal junctions to critical limits.

Atmospheric influences, atomic oxygen degradation, and the thermal variations, along with the impact of the magnetic and electric fields, are classified as effects of the neutral space environment. Although significant failures have been caused by the neutral environment, a higher percentage have been caused by the plasma and radiation environments.

Ionized gases with energy levels less than 100 KeV are usually identified as plasmas. The plasma environment surrounding the earth varies with altitude and latitude but is also heavily influenced by solar activity. In geosynchronous orbits (GEO), spacecraft are bathed in a low-concentration, high-energy plasma that is highly sensitive to solar storm activity. Spacecraft moving through this plasma environment can accumulate differential charges. Arcing can result, overloading electrical components, or exposing surfaces to further damage. High-energy particles can also penetrate insulating material, causing leakage paths in electrical networks. Spacecraft charging has been a cause of many significant failures, most notably in GEO communication satellites.

The radiation environment is perhaps the most significant in terms of spacecraft failures. The radiation environment is characterized as containing energetic particles (ranging from 100 KeV up to several GeV) that are either trapped by, or passing through, the earth's magnetosphere. This radiation takes the form of cosmic ray particles, solar protons and heavy ions, and fast electrons. These energetic particles readily penetrate a spacecraft's shell, displacing materials at the atomic level. They unusually have an immediate effect if they happen to impact an electronic component.

How radiation affects circuitry depends of the type and energy of the particle. The majority of occurrences involving radiation are single-event upsets (SEUs), which cause a state change, such as a digit being "flipped" from a zero state to a one. Such events are common and not of concern in most circumstances, since error detection and correction (EDAC) software can locate and reverse the event. Another type of event, of considerably greater concern, is the single-event latchup (SEL), which causes a part to draw excessive current until it is shut down. SELs are serious, but cycling the power to a component will usually reset a circuit. Corollary damage can result, however, since the SEL is effectively a short circuit. The temporary short circuit can overload the power supply, or reduce bus voltage, damaging power-sensitive electronics. A third type of failure is the single-event burnout (SEB). An SEB is not recoverable.

The sensitivity of equipment to radiation damage is a particularly important issue in relation to commercial plastic components, which are used extensively in small spacecraft. This is covered in more detail in Appendix D.

Design Failures

Engineers designing spacecraft components now build upon decades of measurement of, and experience in, the space environment. Yet design failures remain a major source of failure. A design failure occurs when the strength of parts or components, purchased or manufactured, proves insufficient to withstand the loads experienced during the mission. If the load experienced was associated with a phenomenon not yet understood, or of a magnitude not yet recorded, the failure is usually assigned to environmental causes. Design failures are, therefore, associated with oversight or error. It is worth noting that there is no category for failures related to the use of new technology. Failure of a new design would fall under the category of a design failure.

Failures Related to Parts and Quality

When a failure occurs in the absence of unexpected environmental loads or a clear design error, it is usually classified as being caused by a quality or parts problem. Parts failures usually occur randomly. Spacecraft rarely carry sufficient instrumentation to identify the failure of a discrete part. Instead, a component or collection of parts is identified as the source of failure. Failures related to quality occur when similar parts have repeated problems or when a ground test reveals weaknesses in a representative sample of similar parts. Failures caused by incomplete testing, or induced by testing, are also classified as related to quality.

It is important to realize that component reliability data often assume ideal handling and processing. In commercial settings, components are sometimes never touched by human hands. Entire systems are assembled by automated processing equipment. Spacecraft applications remain custom in the sense that human technicians handle parts and components throughout the process. The effects of part handling and processing on manufacturer predictions of reliability are not well understood.

Other Types of Failures

Many failures can be traced to a variety of other events. Operational errors account for some reported failures. These events are usually related to human error, in which a ground operator issues a command that overloads a spacecraft system or component or exposes sensors or instruments to out-of-bound conditions. Many anomaly databases classify normal aging, wear, or depletion of consumables as a failure. Software-related problems are also classified as "other" failures.

Spacecraft Failures Over Time

Unlike in terrestrial systems, where strain and wear cause failure rate to be a linear function of time, spacecraft failure rates diminish over time. This relationship, a Weibull distribution, is shown in Figure B.1, where the average number of failures reported annually for a given spacecraft decreases.

Analyses that do not account for the relationship shown in Figure B.1 will usually produce overly pessimistic reliability and lifetime estimates. A pessimistic reliability estimate can lead to additional design effort and biased performance trades. This may lead to a more robust design and ultimately to a more reliable spacecraft—which is not always desirable, because additional design efforts usually increase TMC.

Spacecraft Failure Trends

The categories outlined above are sufficiently common among failure databases to get an idea of where spacecraft reliability is headed. These categories are used in Figure B.2 to draw some high-level conclusions about failure.

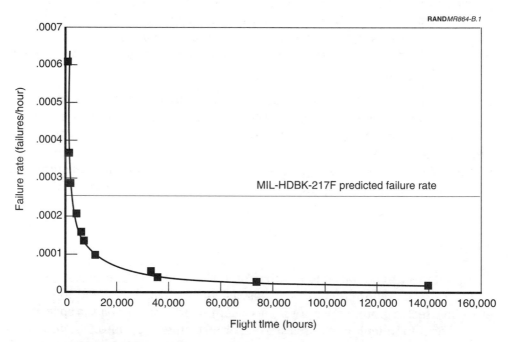

SOURCE: Krasich (1995), p. 3.

Figure B.1—Decreasing Rate of Failure with Time on Orbit—Voyager Spacecraft

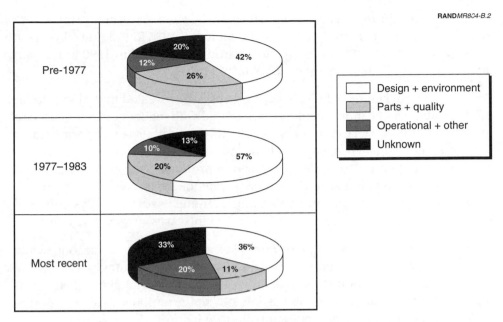

NOTE: This chart excludes data from the SAMPEX mission.

SOURCES: Pre- and 1977–1983 data, Hecht et al. (1985), p. 47; most recent data, Remez et al. (1996), p. 33.

Figure B.2—Spacecraft Failure Trends

Design and environment causes continue to be the most significant sources of failure. The reductions shown in the most recent data are possibly due to improved design techniques and the use of more refined environmental models. Yet the fact that design and environment factors remain the largest single cause of spacecraft failure is alarming. Failures caused by attributes of the space environment that have been documented and characterized are usually classified as design failures. Since our knowledge of the space environment has steadily improved, it can be presumed that mistakes, or insufficient design margins, are a major barrier to further reducing failure rates. Another recent estimate placed design errors at the top of the list of spacecraft failure causes. (Fleeter, 1997a, p. 14.) Design errors are also predominant in failures that occur prior to launch. A recent review of planetary spacecraft noted that 60 percent of the failures that occurred during test and integration could be traced to design problems. (Gindorf, et al., 1994c, p. 12.)

Parts have traditionally been viewed as the source of failure in spacecraft systems. In the early years of the space program, this was indeed the case, mainly because of quality and reliability problems with evolving microelectronics. Yet recent data clearly show that parts and quality factors are the minor constituent of spacecraft failures. These data are corroborated by other studies. An analysis of failures in aircraft avionics conducted in 1971 found that 50 percent could be traced to part failures. A similar study conducted in 1990 found part failures to be negligible. (Pecht et al., 1992, p. 1161.)

Scientific investigations of failure mechanisms have revealed that many claims of part failure can be more accurately associated with inadequacies in design or improper handling of components. A recent JPL study reviewed parts-related failures in the Viking, Voyager, Magellan, and Galileo spacecraft. Only 27 failure reports for these missions could be traced to problems with parts, and all but eight were later attributed to design or test deficiencies. None of the parts-related problems were considered serious, although redundant systems prevented an escalation of the problem in seven of the cases. (Gonzalez, 1996b.)

Since the category "other" contains normal wear and old-age events, one would expect this percentage to grow over time. The lower percentage in the 1977–1983 period is possibly due to spacecraft beginning to live longer in this reporting period and to the use of more-sophisticated ground control techniques, which reduced the number of operator-induced failures.

Failure Effects

Another aspect of spacecraft failure data is the severity of failures when they do occur. Figure B.3 depicts the trend in reported failures. The "pre-1977" and "1977 to 1983" data were based on a long-term study of approximately 300 spacecraft. Approximately 36 percent of the early failures were significant; in the 1977 to 1983 sample, the ratio had dropped to 19 percent.[4] A recent study by the NASA Goddard Space Flight Center of 21 spacecraft revealed only 112 anomaly reports with only three significant incidents.[5] (Remez et al., 1996.)

Properly designed spacecraft systems can withstand a myriad of component failures and operating anomalies without suffering a significant loss in performance. The Voyager 1 and 2 missions are among the most notable examples. The Voyager program, widely recognized as a hallmark in planetary exploration, dealt with many component problems throughout its long flight history. (Gonzalez, 1996b.)

---

[4]Both the pre-1977 and the 1977–1983 data are presented in Hecht et al. (1988), p. 14.

[5]The 112 reported incidents exclude an additional 100 incidents from the Small Explorer SAMPEX spacecraft. The SAMPEX data are excluded because the Small Explorer program collects failure-mode data in a form different from other GSFC offices.

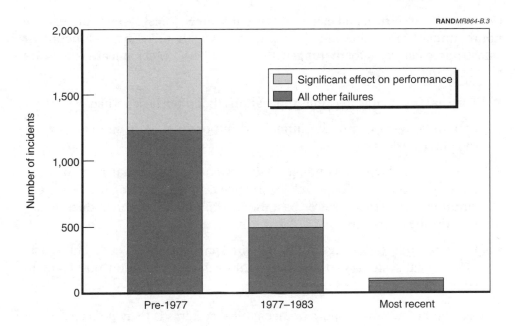

Figure B.3—Decreasing Severity of Failures

Failure in Mechanical Systems

In terms of failures onboard spacecraft, perhaps the area of greatest concern is the performance of mechanical systems. The performance and reliability of electrical and electronic components have improved dramatically in recent years. The design and development of mechanical systems, however, have not advanced in parallel. Many of the most serious recent spacecraft anomalies can be traced to mechanical system failures, as outlined in Table B.1.

Table B.1

Examples of Mechanical Failures in Recent Spacecraft

Mission	Event	Impact	Likely Failure Mode
Mars Observer	Propulsion system failure	Loss of Spacecraft	Leakage and ignition of hypergolic propellants—rupture of high-pressure lines
Galileo	Stuck high-gain antenna	Degraded performance	Excessive friction due to misalignment in antenna restraint pins
Alexis	Damaged solar array	Degraded performance	Attachment bracket broke free after deployment
Mars Global Surveyor	Failure to latch solar array	Modification of flight plan	Structural failure of solar array damper arm attach fitting

Compared to electrical and electronic systems, mechanical systems are usually nonredundant; when they fail, there is greater likelihood of loss of function or catastrophic failure. (Oberhettinger, 1994, pp. 16–24.) Mechanical systems are unique in that

- They are frequently first-time applications that often lack heritage.

- Repetitive testing is often difficult or impossible, as in the case of pyrotechnic devices.

- It is more difficult to conduct environmental testing that recreates the forces a mechanical design will experience in space. Testing in a one-g environment can stress devices past the design point, possibly inducing failures during operation.

- Long periods of storage or transit in space often precede their use. Mechanical systems can lose lubricant or gather corrosion that leads to later failure.

One method of avoiding failure in mechanical systems is to avoid using them. Future missions will likely require more complex mechanical systems, however, so avoidance will not be a reasonable approach for most missions. Advanced mission concepts, like deployable structures, will require miniature mechanical devices that are both reliable and precise. Improving the reliability of mechanical systems remains, therefore, a high-priority item.

RESEARCHING FAILURE MECHANISMS—THE PHYSICS-OF-FAILURE APPROACH

In the modern marketplace, quality and reliability are more closely related than ever before. To remain competitive, product manufacturers have applied ever more stringent quality standards, enabling them to deliver higher-reliability devices.

Underlying the drive for better quality and reliability is a shift from empirical understandings of failure mechanisms to a more scientific approach. The *physics-of-failure* approach applies reliability models, built from exhaustive failure analysis and analytical modeling, to environments in which empirical models have long been the rule. (See Pecht, 1996b, and Stadterman et al., 1996.)

Scientific approaches to failure are certainly not new. The physics-of-failure approach embodies techniques well known to structural engineers responsible for building large structures—only one unit is built, and a failure would mean significant loss of life and property. The central advantage of the physics-of-failure approach is that it provides a foundation upon which to *predict* how a new design will behave under given conditions, an appealing feature for small spacecraft engineers.

In terrestrial applications, the physics-of-failure approach has helped to increase design confidence and boost quality and reliability, perhaps most readily demonstrated in the case of microelectronic components. In terms of quality, the defect rate for high-volume electronics, for example, is now so low that traditional methods of acceptance testing make little sense.[6] The process that produces Intel's Pentium® microprocessor, a complex device with fine feature sizes, averages 17 defects per million units produced.[7] Strategies that identify and replicate proven components, such as Known Good Die (KGD) practices, are also reducing defect rates in highly integrated (stacked) electronic components.[8] Reliability improvements are equally impressive. The Pentium® processor has a mean time between failures (MTBF) of 36 million hours. This level of reliability means that the central processing unit (CPU) is unlikely to fail within the normal lifetime of a modern personal computer. The high quality of mass-produced microelectronic components has led the automotive sector to set some of the industry's most stringent qualification standards, with zero allowable rejects (Pecht, 1996c, p. 22).

In space applications, the physics-of-failure approach seeks to augment the traditional postmortem analysis of failure data with an expanded knowledge base of failure mechanisms. Data from physics-of-failure research should help to reduce both failure rates and failure severity, improving the reliability of space components and systems. The physics-of-failure approach could also assist with

- *Application of technology*. More so than in the past, small spacecraft rely on advanced technology for which few historical reliability data are available. The physics-of-failure approach offers a means of evaluating how new designs will operate, based on a more refined understanding of the response of materials and the behavior of analogous systems.

---

[6]Research conducted at Rome Air Force Base's Reliability Analysis Center (RAC) in 1992 concluded that the average failure rate for commercial electronics was approximately 0.02 failures/10^6 hours (see Priore and Farrell, 1992).

[7]Quality data for Intel microelectronics are available at its developer Web site: **http://support.intel.com/oem_developer/**.

[8]Dense electronics, such as the stacking of chips into multichip modules (MCMs), take up less space, require less power, and are easier to integrate than discrete components. These attributes make MCMs very popular among small spacecraft builders. MCMs typically cannot be tested before final fabrication, however, at which point a chip fault requires scrapping the part. To help prevent chip failures, integrated circuit dies that are known to be error free are precertified—the KGD process.

- *Risk mitigation.* Gains in robustness and component reliability translate directly into reduced overall mission risk. Also, an ability to estimate reliability at the part or component level supports improved design optimization of the overall spacecraft.

- *Failure awareness.* The physics-of-failure methodology allows the spacecraft engineer to predict with greater assurance the "first failure" of a given component or design.[9] By creating new tools for predicting performance, the physics-of-failure approach can help spacecraft teams focus on aspects of failure analysis early in the design process.

The study of failure physics, like any other scientific discipline, requires testing to validate hypotheses and gather data on failure mechanisms. A significant amount of research can be conducted on the ground, but some amount of space-based research will likely be necessary. Physics-of-failure research will likely make extensive use of low-cost "time-in-space" facilities, such as Shuttle-deployed free-flying spacecraft. Inexpensive long-duration missions might allow data to be gathered on actual performance in space, with components being returned to earth for analysis.

IMPLICATIONS FOR FUTURE SPACECRAFT

As discussed above, spacecraft have been steadily becoming more reliable. Discounting cases of catastrophic loss, failures have been generally fewer and less likely to affect mission objectives significantly. It is, however, unclear that this trend will continue with the current generation of small spacecraft.

Chapter Two mentioned that one of the consequences of budget reduction and the shift to smaller spacecraft has been a greater willingness to accept risk. Later, Chapter Five described the "risk as a resource" approach, in which risk is treated as a variable in the many engineering trades that are made during the planning and design of a mission. Created from discussions with small spacecraft teams, Table B.2 presents a qualitative assessment of how many of the steps taken to achieve cost reduction affect risk.

Some small spacecraft trends raise the potential for failure, introduce new failure sources, cause failure sources to be overlooked, or reduce the spacecraft's resilience. Perhaps the most obvious example of this is the potential for launch

---

[9]For terrestrial applications, component manufacturers are concerned primarily with the number of failures likely to occur in a given period of time (failures in time or FIT), or the mean time between failures (MTBF). These data are less useful to the spacecraft engineer. For commercial components, increasingly used in small spacecraft, the importance of the problem is elevated, since manufacturers cannot, and often will not, supply the information a spacecraft designer needs to ensure design reliability (see Appendix D).

Table B.2

Risk in Small Spacecraft Programs

Small Spacecraft Strategy	Manifestations	Risk of Failure
Simplified design	Rescoping mission requirements	Neutral
	Design reuse	Decrease
	Reduced redundancy	Increase
	Mainly incremental improvements	Neutral
	Use of commercial plastic-encapsulated electronic parts	Increase
Streamlined test procedures	Test at higher level of integration	Increase
	Reduced test plans	Increase
Reduced procurement oversight	PI-mode management	Neutral
	Performance-based contracting	Neutral
Other attributes	Smaller teams	Increase
	Small launch vehicle	Increase

failure. Small spacecraft operate in a weight class in which there is a significant risk of failure associated with the launcher. Of the four Air Force STEP spacecraft that have been completed, two have been lost due to launch failures. Several new small launch systems are currently being developed, but it is unlikely that they will be able to demonstrate near-term reliability.[10]

Lengthy launch delays can also introduce sources of failure. Unplanned periods of dormancy can lead to such failure-inducing situations as lubricant loss, the introduction of corrosion, or the loss of battery potency.[11] Small programs are particularly susceptible, since funds may not be available for adequate retesting prior to a delayed launch.

Another area where failure could place smaller spacecraft at greater risk is the use of redundant systems. Failures do not necessarily place mission objectives in jeopardy if backup systems are available or if the spacecraft design is sufficiently flexible to allow workarounds. Historically, redundancy has been a central method of achieving resistance to failure and has been incorporated up to the point at which the incremental costs of including it began to exceed reductions in the cost of failure.

---

[10]Historically, it has taken an average of 57 flights for a new launch system to reach a sustained reliability of better than 75 percent; see Chow (1993), p. 44.

[11]The effects of dormancy on component degradation were studied extensively under the Spacecraft Aging Study at the Air Force Phillips Laboratory using the P-80 (Teal Ruby) spacecraft, which was built but never flown. Loss of lubricant, which occurred during transportation and dormant storage, is suspected to be a major factor in the failure of the Galileo spacecraft to deploy its high-gain antenna fully.

There is considerable disagreement in the small spacecraft community regarding the use of redundant systems. Many engineers feel that "single-string" systems are inherently reliable because of their simplicity. In general, redundancy increases the complexity of the spacecraft, which is contrary to the notion of reliability through simplicity. Increasing levels of component and device reliability further argue against the need for backup systems. Redundancy is also costly in terms of the resources that must be devoted to backup systems. Redundant systems add mass, consume power, require more wiring, and increase the dimensions of the software used to operate the spacecraft. Cost, however, is the most usually cited reason for limiting the use of redundancy. Many engineers feel that budgets are simply not adequate to consider redundancy in small spacecraft.

Although it adds a financial and technical burden, redundancy has been a critical factor in many successful past missions. As noted above, the Voyager mission experienced many failures, but these were largely countered by redundancy and workarounds. Redundancy has been shown to be especially important in certain systems. The telecommunications system is one example: If ground controllers lose the ability to "talk" to a spacecraft, the potential for inflight repair or reconfiguration is also lost. A 1994 JPL study of the critical telecommunications system on six prior missions (Voyager 1 and 2, Viking 1 and 2, Galileo, and Magellan) revealed that redundancy likely saved five of these missions from catastrophic failure. (Brown, 1994, p. 14.) The affordability of redundancy, in terms of using it on small spacecraft, is also changing. Discussions with component and subsystem suppliers suggest that the cost of adding redundancy is declining. Several small spacecraft missions (for example, Discovery-NEAR and SSTI-Clark) have taken advantage of these trends and implemented designs that are heavily redundant, within tightly constrained budgets.

Offsetting the higher risk of failure associated with such areas as launch and use of redundancy are trends that promise continued reduction in the number and severity of failures. Spacecraft parts and equipment are expected to continue to become more reliable (a subject covered more thoroughly in Appendix D). Future small spacecraft are likely to rely more heavily upon autonomous systems for fault detection, isolation, and recovery. An early example of this trend is the EDAC software now used extensively to correct automatically for SEU errors. Future autonomous software agents, capable of resolving complex problems and reconfiguring spacecraft systems, have the potential to reduce failure effects significantly. (Man, 1997, p. 4.) Software is usually less capable, however, when it comes to dealing with mechanical failures and might actually increase the consequences of failure if corrective actions are implemented that impede the ability of ground controllers to intervene. (Oberhettinger et al., 1994, p. 26.)

Management approaches that emphasize "failure awareness" are important elements of efforts to ensure that reliability improvement trends continue. Noteworthy throughout the course of this study was the variation in approaches taken to manage risk in small spacecraft programs. Risk-management approaches often depend on the experience of the most senior engineers on a given team. Reducing this variability is the major goal of NASA OSMA's effort to formalize risk management approaches under the "risk as a resource" theme. Repeating past mistakes has also been a source of frustration for many programs. This has spurred efforts to pass on the experiences of senior managers and engineers to younger spacecraft designers.[12]

Summary

Almost four decades of experience in building spacecraft and measuring the space environment have yielded a refined understanding of how to avoid failure. This experience is reflected in the fact that

- The number of spacecraft failures has been steadily decreasing.

- Failures, when they do occur, are less severe.

Within those trends, however, are some significant areas of concern that could affect continuous improvement in mission performance:

- Design-related failures are playing a more significant role as the total number of failures diminishes.

- Mechanical failures contribute significantly to reduced performance or loss of spacecraft.

These areas deserve special attention in terms of focusing failure-reduction initiatives.

An area of great promise, in terms of understanding failure mechanisms and improving the construction of spacecraft, is the physics-of-failure approach. The goal of this approach is to replace empirical models of failure with more rigorous scientific analyses of how failure occurs in spacecraft components and subsystems. The increasing accuracy of failure models should aid in reducing

---

[12]NASA GSFC has consolidated past experiences into a Space Engineering Lessons Learned (SELL) database. JPL sponsors a "Common Threads" workshop to relate past experience; see Brown et al. (1996).

the number of design errors and serious mechanical failures. The physics-of-failure approach is also important in terms of helping to predict the performance of new technology and providing new tools that increase an awareness of failure early in the design process.

To assist physics-of-failure initiatives, a greater degree of cooperation between the various organizations collecting and disseminating failure data is needed. The adoption of common recording and reporting formats would assist in the preparation of actuarial data. Funding for joint analysis efforts might be considered with the aim of providing a foundation for improved reliability and longevity estimates.

NASA has adopted a higher-risk approach in shifting to smaller spacecraft. One possible outcome is, in the short-term, a higher rate of failure that would disrupt the trends described above. Yet, as the reliability of small launchers improves, as small spacecraft programs incorporate high reliability systems, and as new design techniques proliferate, it is likely that future small spacecraft will continue the trend toward fewer failures.

TESTING FOR RELIABILITY IN SPACE SYSTEMS

INTRODUCTION

Testing is a pivotal phase in the development of space systems. Although it is an ongoing process, which actually begins with the development of test plans early in the design phase, the most crucial time is the period of full-up operation of the integrated system. A barrage of tests, often as many as a program can afford, is applied to the final product in an all-out effort to eliminate defects undetected in earlier inspections and operations.

Improving the process of testing has broad implications for risk reduction in space systems. Better test procedures affect not only the ability to build more reliable spacecraft but also the launch and the ground control systems that deploy and operate them. Since change in this area has such far-reaching implications, efforts to improve testing are very important.

Traditional methods of testing spacecraft are undergoing a period of change, however. This appendix will review the factors driving change within the test community, highlight NASA's response, and assess the impact on space systems.

FACTORS DRIVING CHANGE IN THE TEST COMMUNITY

The process of testing spacecraft, much like the process of designing them, is being significantly restructured. Three factors are driving this restructuring: a policy decision to deemphasize the role of military specifications; new data that are challenging traditional test methods; and the emergence of small spacecraft that demand new, less costly approaches.

Moving Away from Military Specifications and Standards

Change within the test community intensified in the summer of 1994 when then–Secretary of Defense Perry issued a memorandum restricting the use of military specifications and standards (milspecs and milstds) in DoD procurement practices:

> Performance specifications shall be used when purchasing new systems, major modifications, upgrades to current systems, and non-developmental and commercial items, for programs in any acquisition category. If it is not practicable to use a performance specification, a non-government standard shall be used. (Perry, 1994.)

Secretary Perry's decision allows the use of milspec only "as a last resort." Although the decision focused on the procurement of military systems and not on the use of milspecs for design and test, it nevertheless sent shock waves through the PA community, which relied heavily on milspecs for source material. (Coppola, 1995.) In principle, the announcement signaled a movement away from tradition and, over the long term, the possibility that the milspec backbone of the test community would dissolve.

The spacecraft community makes extensive use of milspecs for design, test, and inspection. Through decades of use, they have become embedded in the PA processes of both the government and the aerospace industry. The Perry decision was viewed with some disappointment within the PA community, since more recent milspecs were created to perform a dual-use function and had been significantly streamlined. New standards for quality and reliability, such as ISO-9000 and the proposed program management standard, ISO-14300, are viewed by many spacecraft developers more as guidelines than specifications and are not, therefore, seen as a replacement for traditional milspecs.

New Insights into Traditional Test Practices

The physics-of-failure approach, outlined in Appendix B, is beginning to illuminate deficiencies in traditional test practices and inspire new approaches to testing. It appears that some tried-and-true test practices can be ineffective when applied to new technologies, possibly even inducing failures. There is some evidence, for example, that burn-in tests, the principal means of identifying infant mortality in new parts and electronic components, can actually lead to premature system failure. (Jordan et al., 1996, p. 18.)

Another example is thermal testing. Repetitive thermal cycling, where systems are alternately heated and cooled inside a chamber, is used extensively in spacecraft programs. Analysis of thermal-cycle testing now suggests that a single thermal excursion is sufficient to identify most defects and that additional variations can induce failure. (JPL, 1994, p. 14.) The traditional practice of derating components (operating well below the stated performance limit of a part or component or, conversely, selecting a component with performance well above expected stress levels) in an effort to increase reliability, has been shown to be ineffective in some cases. (Pecht et al., 1992, p. 1163; Coppola, 1995.) The PA community must rapidly respond to these developments and provide guidance to ongoing programs.

Simulation-based design (discussed in greater detail in Appendix E) is also significantly changing the type and level of testing being contemplated for future programs. The ability to simulate the performance of a proposed design with ever-increasing fidelity changes the manner and extent of testing of the final item. It does not, however, replace it.

Small Spacecraft Test Requirements

Although similar to traditional spacecraft, the new generation of smaller ones places additional requirements on the test community by

* Placing a greater emphasis on the cost-effectiveness of testing.

* Elevating the criticality of testing.

* Utilizing new technology to a greater extent than previous missions.

Small spacecraft demand a cost-effective test program. The cost of time in test facilities has not shrunk in proportion to the size of the spacecraft. The equipment used to test a small spacecraft, such as a thermal-vacuum chamber, is often the same equipment as used for larger spacecraft. Designing an affordable test program is, therefore, a challenge for small spacecraft program managers.

Testing, to a significant degree, is a "seat-of-the-pants" affair in small spacecraft programs, and approaches vary widely. As shown in Figure A.1, the test-and-integration phase of NASA small spacecraft reviewed in this study accounts for an average 3 percent of TMC. But this value ranges from 1.3 to 5.5 percent, possibly reflecting the variation in approaches taken to testing. The process of creating a test plan, tailored to both the mission and available funds, relies heavily on the experience of the senior team members. Many of the data validating test procedures are empirical and anecdotal, complicating the task of preparing the most cost-effective solution to the test requirements of a given mission. Since the test phase occurs late in the development cycle, even a carefully crafted test plan can be stressed by cost overruns in other areas.

Testing is more critical in small spacecraft programs, since, in many cases, redundancy is available only partially or not at all. In these situations, testing becomes the most important means of evaluating the design and preventing defects that now have a higher potential for causing failure later in the mission.

To achieve higher performance, small spacecraft rely on technology more than did past missions. Designers employing new technology require performance estimates and projections of risk. New parts and components with little or no space heritage are particularly vexing in terms of test and validation. The closest analog often serves as the only model for designing the test plan.

Additionally, this new technology increasingly takes the form of commercial-grade electronics that are proving to be at least as good as their military-grade equivalents in terms of quality.[1] Their use, however, adds additional complexity in that established methods may be wholly inappropriate to their handling, processing, and test. The screening of electronic parts is a case in point. Screening is a process of accelerated testing designed to eliminate infant mortality. The process helps to ensure that, once in space, parts fail only due to the inevitable degradation of the space environment. Because past component quality was low and the cost of failure high, NASA screened nearly 100 percent of the electronic components used in spacecraft. Detailed procedures for inspection and test were outlined in milspecs and NASA handbooks. For the reasons cited above, however, new test methods are needed.

DEVELOPING NEW APPROACHES TO TESTING SPACECRAFT SYSTEMS

Under the sponsorship of NASA's OSMA, research on test effectiveness has been under way since 1992. Similar efforts are being supported by the Air Force. The advent of small spacecraft has accelerated these efforts, however, and given emphasis to reducing the cost of implementing test procedures.

Figure C.1 portrays the spacecraft testing process. As illustrated in the figure, the goal of each test is to halt defects that have "escaped" detection in prior tests. Occasionally, the test process fails to detect a certain mode of failure, and a defect may make it through the entire test process. Of course undetected defects do not automatically lead to a mission failure; robust designs and redundant systems can protect spacecraft from serious, often compounding, failures.

The width of the rectangles in Figure C.1 implies that some tests are highly effective at detecting certain types of defects; others are more broadly able to find flaws and errors. Additionally, Figure C.1 suggests diminishing utility of additional testing if prior evaluations have already eliminated defects.

---

[1]The use of commercial components in spacecraft is complex and controversial. Appendix D covers this issue in greater depth.

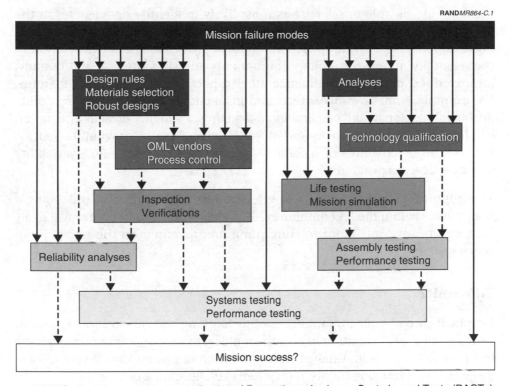

RAND*MR864-C.1*

NOTES: Each box represents a collection of Preventions, Analyses, Controls, and Tests (PACTs). Dotted lines represent "escapes"—undetected/unprevented failure modes.

SOURCE: JPL (Cornford, 1996a, p.4).

Figure C.1—Defect Propagation Model

The goal of NASA's test effectiveness program is to provide improved test procedures that cost less to run and that have a greater probability of eliminating defects. To achieve this goal PA engineers are conducting controlled experiments on current test procedures to evaluate how well they identify defects. This analytical approach has the added benefit of removing past empiricism and providing a better characterization of any given test. Spacecraft builders will have an improved awareness of the impact of specific tests and their limitations.

IMPLICATIONS FOR FUTURE SPACECRAFT

Better characterization of tests allows builders of space systems to tailor a test plan more precisely to the type of spacecraft and mission. A mission that would, for example, place a spacecraft in a high or unique radiation environment should be able to select tests most likely to identify deficiencies in the spacecraft's ability to withstand such conditions. More effective test procedures should also provide spacecraft builders greater assurance in relation to the reliability of new technology components and subsystems. Additionally, improved test procedures influence other aspects of building space systems. For example, some in-process tests and inspections, conducted at key points during fabrication, could be eliminated if testing at a higher level of integration can be shown to be less costly and more effective. Finally, providing scientific data on the effectiveness of specific test procedures should reduce variability among spacecraft programs.

The importance of improving test effectiveness has resulted in this being a major focus within the PA community. More-refined test procedures will likely be a significant contributor to continuing improvement in the reliability of space systems.

SUMMARY

Despite its broad-scale importance, testing has long been an empirical process, with a great deal of variation in how the builders of space systems approach the test phase of a project. Change within the test community has begun as new insights into how components and systems fail illuminate deficiencies in traditional practices. The inherent quality and reliability of new components has also spurred a reexamination of test procedures. This review was accelerated by the government's decision to emphasize performance specifications in DoD procurement over long-standing military specifications and standards. Within the PA community, this decision was viewed with alarm in that milspecs form the backbone of traditional test procedures. Finally, the arrival of smaller spacecraft placed unique demands on test procedures: Testing had to remain effective yet be responsive to a smaller, tightly constrained budget; its criticality was increased, since many small spacecraft were proceeding with little or no redundancy; and a greater amount of new technology was being incorporated.

The response from the PA community has been a fresh look at test procedures aimed at replacing empiricism with experimental evaluations of the effectiveness of specific tests. Both NASA and the Air Force plan to establish refined test procedures that both cost less to implement and are more effective in preventing defect propagation. The net effect of these improvements is expected to be a significant improvement in the reliability of all space systems.

HIGH-RELIABILITY SYSTEMS

INTRODUCTION

Although the devices used in everyday life are getting ever more complex, they are also becoming extremely reliable. In automobiles, for example, the days of finicky engines and short-circuiting electrical systems are rapidly disappearing. Reliability of something as ubiquitous as a mobile telephone is taken for granted. This is an age of extraordinarily reliable systems.

Appendix B discussed the fact that spacecraft too have been becoming more reliable. Earlier, it was suggested that the shift from risk avoidance to managed risk might disturb the trend toward higher reliability. A focus on high-reliability space systems could, however, reestablish trends. This appendix will review factors associated with the reliability of systems in space and will examine terrestrial analogs. It will also discuss an important aspect of reliability: the increasing reliance on commercial electronic parts in spacecraft. The implications for future spacecraft will also be discussed.

HIGH-RELIABILITY SPACE SYSTEMS

NASA and the Air Force have been reviewing the history of spacecraft performance to codify methods that will lead to higher-reliability systems.[1] New design processes are incorporating reliability models that are more accurate and integrated into failure-analysis data systems. Component manufacturers are making many of these models available, such as Texas Instruments' Computer-Aided Reliability and Maintainability Analysis (CARMA). Government testing laboratories are making other models available, such as the Air Force

---

[1]An example is JPL's High Reliability/Long Life (HRLL) Systems Initiative, which is attempting to identify design and manufacturing practices that led to exceptional missions, such as Voyager and Magellan. NASA and DoD are cooperating on the development of high-reliability systems (see SMC, n.d.). The NASA-DARPA-Air Force-Army-Navy consortium RELTECH, for example, has been formed to pursue high-reliability, high-density electronic devices.

Rome Laboratory's Reliability Engineer's Toolkit. New design practices, such as the simulation-based design environment, will likely become linked directly to equipment-supplier data systems. Recent studies have demonstrated the feasibility of applying such practices to the development of spacecraft (GSFC, 1996b). This is an important development because it provides an opportunity to assess reliability more accurately and because it more readily allows designers to predict how new technology and untested designs will perform.

Attaining new levels of reliability is also of keen interest to the commercial space sector. Future Ka-band commercial satellites will be much larger and more powerful than today's units. To compete with terrestrial service providers, satellite companies must provide quality and reliability equivalent to that of fiber-optic networks. The investment in component and subsystem technology to achieve this level of reliability will likely spill over to developers of science spacecraft.

LONGER-LIVED SPACECRAFT

High reliability does not necessarily equate with spacecraft longevity. A spacecraft designed to operate for one year can meet all of its stated objectives, fail on the 366th day, and be considered 100-percent reliable. Small spacecraft are usually designed for specific missions for which longevity may be neither desired nor cost-effective.

Yet spacecraft that are designed with an emphasis on reliability tend to perform well beyond their original design points. The two Voyager spacecraft were built for a 5-year lifetime. Engineers now believe that both spacecraft, which were launched in 1977, will be returning useful data past 2015. The International Cometary Explorer (ICE), launched in August 1978 with a planned 2-year life expectancy, remained operational until a serious failure ended its scientific career in 1995. Another Explorer, the International Ultraviolet Explorer (IUE), also launched in 1978 with a 3- to 5-year life expectancy, was operating around the clock up until it was shut down on September 30, 1996.

Inexpensive small spacecraft have also proven resilient. Alexis, built by AeroAstro, Inc., was designed to operate for 6 months and has so far lasted more than four years. A recent JPL workshop concluded that planetary missions could be extended beyond 25 years. (Gindorf et al., 1994a.)

There are some good reasons to consider longevity a desirable trait for future small spacecraft:

- *New approaches to space science.* Spacecraft lifetimes that can be extended greatly at modest cost might invite new scientific approaches. It could prove cost-effective to service small observatories in LEO, reconfiguring them for new missions. Extended-mission spacecraft could be situated so as to conduct a sequence of investigations, with each successive campaign modified by the previous results. Greater coordination among programs should also be possible.

- *Unexpected results.* Great discoveries have occurred during the "bonus" phase of past missions. Voyager 2's exploration of the planets Uranus and Neptune revolutionized space science, as did magnetic field data that provided scientists the first detailed survey of the heliopause.

- *Unexpected opportunities.* Scientific directions are rarely predictable, and important natural phenomena occasionally appear with little or no warning. Each long-lived spacecraft contributes to a space-based research infrastructure that can be pressed into service when unexpected events occur.

- *Secondary science functions.* Spacecraft can be built to perform a variety of long-term investigations following the completion of the primary mission. The ICE spacecraft, for example, was able to support the European Ulysses spacecraft in building detailed maps of the heliosphere.

- *Planning flexibility.* Coordination of missions that rely on data continuity or service overlap is presently very challenging because of uncertain budgets and shifting priorities. Longer-lived spacecraft can help ensure availability of data.

- *Training.* Spacecraft that live beyond their primary and secondary mission phases become useful tools for training. The utility of spacecraft for education might extend beyond the training of engineers and technicians, to college and high school students. Science spacecraft can also serve as a source of inspiration for K–12 students. (Siewert, 1996, pp. 3, 26.)

Long-lived spacecraft create something of a dilemma in that funding is often not available to continue operation beyond the planned end date. Spacecraft operating requirements are, however, decreasing, and new automated systems should require little human intervention. It should be possible to design future small spacecraft that are also *smart interpreters* in terms of signaling only when a desired phenomenon has been located or when viewing an unexpected event. (Aljabri, 1996, p. 2.)

COMMERCIAL COMPONENTS IN SMALL SPACECRAFT

In 1994, President Clinton signed Executive Order 12931, which mandated that the government, when purchasing equipment, "increase the use of commercially available items where practicable." (White House, 1994, Section 1, Part [d].) This order reflected the increasing awareness within government that equipment and components had reached levels of quality and reliability at least on a par with the requirements outlined in milspecs, complementing Secretary Perry's announcement on abandonment of milspecs for procurement.

Before these 1994 pronouncements, commercial equipment was already an increasingly important element of spacecraft programs, especially small ones. Before discussing this trend and its implications, however, it is important to establish some terms in relation to the use of commercial items in space systems. First, "commercial items" does not refer to government procurement of single or limited numbers of commercially manufactured parts. Rather, the term signifies the use of items, usually parts or components, that are manufactured in large quantity for use in terrestrial commercial applications. The second important consideration is that the term refers to items manufactured to commercial production standards that are usually as good as the practices outlined for decades in milspecs. Commercial parts are sometimes viewed as hobby electronics. In fact, as this appendix will describe, commercial parts used in spacecraft meet extremely high standards. High-grade commercial parts are thoroughly tested, and test data and information about the manufacturing process are usually available.

The use of commercial parts is related mainly to electrical, electronic, and electromechanical components (EEE parts), so the following discussion will deal mainly with this aspect.[2] Price is not the main reason for the increasing use of commercial EEE parts. As shown in Figure A.1, parts average 2 percent of TMC. Spacecraft designers have turned to commercial parts because they offer quality and reliability, but more importantly, they are increasingly the only option in terms of availability and performance.

Quality and Reliability of Commercial Parts

Military and space applications have always a placed high demand on the quality of parts and components. In the early days of the space program, performance requirements exceeded the manufacturing capability of the electronics industry. Military-grade microelectronics were typically encapsulated in ceramic to prevent moisture from reaching circuits and to "ruggedize" circuitry from shock and radiation. At the time, variations in manufacturing techniques and quality control practices among component suppliers led to unacceptably high defects-per-million (DPM) counts.

---

[2]The Workshop Notes of the EIA/IEEE/ISHM Joint Meeting on Electronic Components for the Commercialization of Military and Space Systems, San Diego, Calif: February 3, 1997, provides a detailed review of the status of commercial electronic parts.

To ensure the quality of incoming parts, the government began screening electronic components in the 60s according to procedures defined in MIL-STD-883, Quality and Reliability Assurance Procedures for Monolithic Microcircuits, a document still in use today. In the 70s, additional milspecs were drafted to define requirements for "space-rated," or Class-S, components.[3] This environmental-stress screening (ESS) effectively weeded out substandard components, and screened parts became the building blocks of spacecraft systems. Parts-related problems dropped dramatically.

Automotive, consumer electronic, and machine-tool quality and reliability requirements in many cases matched those in the aerospace field, but they added a new dimension—a need for low cost. Electronics manufacturers responded by heavily automating their processes (which dramatically improved quality) and switching to plastic-encapsulated microcircuits (PEMs), the form most common today. The cost of electronics plummeted. In 1995, for example, a Texas Instruments digital signal processor cost $947 encapsulated in military-grade ceramic, $400 in military-grade plastic, $182 in commercial-grade ceramic, and $73 in commercial-grade plastic.[4]

From the creation of the industry, both the quality and the reliability of commercial electronics have improved steadily. Failure rates for highly integrated electronics, such as the x86 family of Intel microprocessors (including the Pentium® and Pentium Pro® designs) have followed a similar path, as shown in Figure D.1 (Intel, 1996, pp. 1–3). Here, failure rate, also called FIT, is plotted alongside DPM.[5]

Today, however, unscreened commercial-grade electronic components are demonstrating reliability matching that of screened Class-S parts. Figure D.2 compares integrated circuit (IC) failure rates (measured in occurrences per million hours of operation) of unscreened commercial-grade items with Class-B and Class-S screened components (Plum, 1990).[6]

---

[3]A thorough review of past and current government practices for designing and testing electronic parts, as well as a case for restructuring these practices, can be found in Pecht (1996a).

[4]Results of U.S. Army study as quoted in Pecht (1996a), p. 8. Note that ceramic parts are also used commercially, mainly in high-power applications.

[5]It is important to note that measurements of production quality and reliability may not reflect the figures obtained in applications where parts are handled and assembled by human fabricators.

[6]The analysis used in this essay was based on the controversial Military Handbook 217 and should be considered conservative. MIL-HDBK-217 predicts failure rates for spacecraft electronics using terrestrial analogs—a constant-stress environment and a resultant exponential failure rate. Data from space missions have shown that failure rates are not as severe as those predicted by these models and that failure rates decrease over time, as shown in Figure B.1. As a result, spacecraft engineers rely on MIL-HDBK-217 mainly for comparative analysis. The most recent edition of MIL-HDBK-217, Rev. F, continues to use exponential failure rates as a basis for prediction. Excessively conservative predictions can lead to design and selection decisions that are unnecessarily costly. See Appendix B of Pecht et al. (1992) for a discussion of the shortcomings of MIL-HDBK-217.

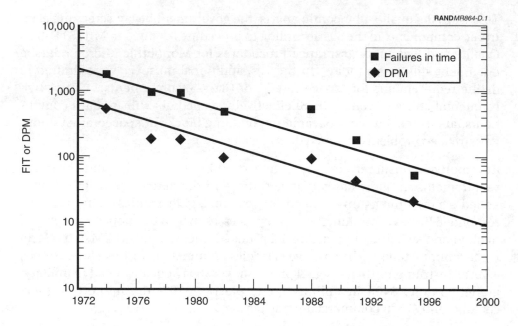

SOURCE: Reprinted by permission of Intel Corporation. Copyright Intel Corporation 1996.

Figure D.1—Intel: Quality and Reliability Trends

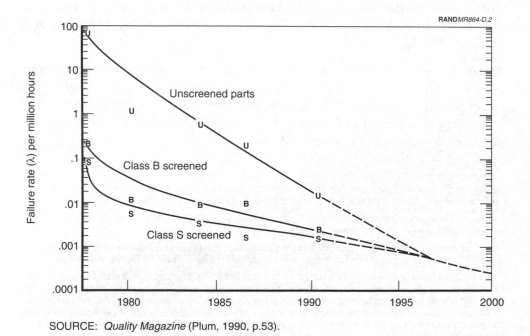

SOURCE: *Quality Magazine* (Plum, 1990, p.53).

Figure D.2—Improvements in IC Performance

As shown earlier in Figure B.2, part- and quality-related failures now constitute approximately 11 percent of reported incidents. Part failures, when they do occur, are believed to be more frequently due to problems in handling and assembling subsystems than to inherent flaws in the quality of the item.

Availability and Performance of Commercial Parts

Sales to the government dominated the early market for semiconductors, but demand from other sectors grew as the power and reliability of components improved. Today, the government represents less than 2 percent of the domestic market (Table D.1). The microelectronics market is characterized by fierce competition; manufacturers must maintain cost-effective production facilities, and many have chosen to bypass the dwindling military marketplace. Among those that have chosen to drop dedicated "military grade" product lines are AMD, Motorola, and Intel. For most of these, Executive Order 12931 and Secretary Perry's decision were important factors in the strategic decisions of the firms to cease production. Importantly, however, other manufacturers, such as Texas Instruments and Analog Devices, have chosen to remain in what is still a billion-dollar military market. In other cases, "sunset manufacturers"have emerged that specialize in the production of critically needed components that leading-edge producers would no longer handle.

Component performance is another important factor. Space-rated components typically represent older technology, since manufacturers will first dedicate resources to production of parts for the broader consumer market. In today's small spacecraft, one can still find radiation-hardened Intel 386-class microprocessors. More-advanced radiation-hard processors are available, but it will likely prove increasingly difficult to keep pace with the growth of commercial technology. Spacecraft engineers are being driven, therefore, to embrace commercial EEE parts, finding ways to ensure proper function in what is arguably the harshest environment of all.

Table D.1

Domestic End Use Demand for Semiconductors

Sector	1970	1986	1996
Communications	16	12	15.7
Consumer	15	6	5.8
Industrial	8	16	8.9
Auto	3	11	6
Computer	12	35	61.7
Government	46	20	1.9

SOURCE: Semiconductor Industry Association (1996). Used with permission.

Spacecraft Application of Commercial Parts

The use of commercial parts in spacecraft has spawned debate and research within the community, much of it focusing on the use of PEMs as replacement for military and space-grade ceramic encased parts.[7] Attention centered around fears that commercial parts could not

- provide adequate resistance to the effects of ionizing radiation

- survive the rigors of testing and subsequent launch

- demonstrate the reliability demanded of aerospace ceramic parts

- survive long-term storage in the period prior to launch or during the long cruise period on deep-space missions.

PEMs have proven to be extremely reliable in demanding applications. The automobile industry, for example, has made extensive use of plastic parts in harsh environments. In 1995, Delco, a major supplier of automotive electronics (8 billion devices in 1995), reported an under-the-hood reliability for microprocessors of approximately 40 failures per million devices in the 5-year, 50,000-mile warranty period, an order-of-magnitude improvement from 1993 (Servais, 1997).[8] The amateur satellite community embraced the use of PEMs early, with a high degree of success. It was such successful applications that invited spacecraft engineers to take advantage of the ready availability of high-performance microelectronics.

For NASA, the migration to commercial EEE parts began in the late 1980s, well before the 1994 executive order. The SAMPEX small spacecraft, built by engineers at NASA GSFC, was a pathfinding mission in many respects. It was an early user of solid-state recorder (SSR) technology, which was built around commercial static random access memory (SRAM).[9] Other data points, such as the Total Ozone Mapping Spectrometer (TOMS) and the Cosmic Ray Upset Experiment (CRUX), confirm that commercial parts provide an acceptable alternative to traditional design options.

Yet the use of commercial PEMs in small spacecraft presents some challenges. Higher levels of integration and greater packaging density on the spacecraft

---

[7]For a detailed account of PEMs, see Pecht et al. (1994). A treatise on the use of PEMs in space applications can be found in Baluck et al. (1995).

[8]Servais's notes provide an excellent compendium of the latest information related to the use of commercial plastic components.

[9]SSRs are now widely used in small spacecraft. Replacing bulky tape recorders, they offer the benefits of lower power, lighter mass, faster access to data, and improved reliability. The performance of the SAMPEX SSR was exemplary, as documented in Seidleck et al. (1995).

mean that PEMs must often survive in a higher-temperature environment. Electronic subassemblies containing PEMs must also be designed to survive an environment usually beyond the shock and thermal boundaries for which components were originally intended. Table D.2 examines some of the pros and cons of using commercial parts in small spacecraft.

The success of missions using commercial EEE parts is a clear indication that engineers have found ways to overcome any shortcomings associated with their use. Research has shown that many of the early concerns, such as component outgassing, are not limitations. Yet some important areas of ongoing research remain.

Remaining Areas of Concern

Radiation tolerance remains a concern for PEMs, despite the fact that discriminate use of non–radiation hardened, commercial-grade components has met with demonstrated success in several applications.[10] SELs can cause serious malfunctions, and PEMs are of concern in regard to their susceptibility to these types of failures.

High-performance complementary metal-oxide silicon (CMOS) parts are typically built with extremely small feature sizes—0.35 µm, for example, in the case of Intel's Pentium® processor. Radiation susceptibility increases as feature size decreases. (See Lauriente et al., 1996, p. 3; LaBel et al., 1996.) A plastic part's resistance to the effects of radiation depends on many factors, including the manufacturing process, materials used, and technology used in its design. Yet many fear that PEMs, operating without the protection of a ceramic shell, will be prone to SEL-type failures or rapid degradation. Additionally, there are some indications that traditional testing methods might render parts more susceptible to radiation later on. (Jordan et al., 1996, p. 16.)

Some of these considerations become amplified in a small spacecraft application. A large spacecraft offers natural shielding, and engineers concentrate sensitive electronics deep within the structure. By its nature, a small spacecraft offers less natural shielding, and the internal organization of subsystems often requires mounting electronics at or near the outer structure. Added shielding for the PEMs adds mass, always a precious resource on a small spacecraft.

---

[10]The risks associated with using non–radiation-hardened electronics depends on the spacecraft application. A memory device, in which a fast proton may temporarily flip a bit, is in a different category than a system controller that might, for example, operate the primary propulsion system. For more information on radiation effects, see Lauriente et al. (1996) and LaBel et al. (1996).

Table D.2

The Pros and Cons of Commercial Parts

Pros

Much less expensive—saving can be passed on to design and test.

Lighter than ceramic components.

State-of-the-art performance, allowing a more robust design and a greater capacity to deal with unexpected failures.

Widespread availability.[a]

In some cases, commercial parts offer greater reliability and resistance to overstress.

Reduce the burden of tracking and qualifying Class-S standards, reducing vendor certification and thus saving time.

More choice of suppliers.

Greater product variety.

Surface mount technology allows manufacture in slimmer packages.

Stringent quality and reliability monitoring.

Commercial users continuously driving for improved performance and better quality and reliability.

Cons

Significant redesign is often necessary. Designers have sometimes found that the cost of modifying commercial components exceeds the purchase cost of space-grade equivalents or the cost of designing custom components.

Quality and reliability data are not easily used in spacecraft design—MTBF is less relevant than "time to first failure."

Operating characteristics (internal wiring specifications, physical dimensions, etc.) are often changed on short notice, mainly in response to the needs of the computer market. Equipment and procedures used to test and integrate these components can quickly become obsolete.

Performance data might not be available, or might be considered proprietary by the manufacturer.

Greater sensitivity to radiation dosages necessitates extensive testing and protection schemes.

Long-term performance of plastic parts in space is uncertain—longevity is a potential issue.

Competition can cause component manufacturers to advertise higher performance for commercial than for equivalent military-grade components.

Parts can suddenly be made obsolete and disappear from the market.[b]

Automated processes are usually used to assemble commercial parts. Spacecraft are hand-built, which increases the potential for mishandling sensitive electronic devices.

Requires more careful assembly.

Is more sensitive to mechanical and thermal stresses.

[a]Due to lack of parts availability, it is not uncommon for subsystem assemblies, such as printed circuit boards, to be delivered for integration into a spacecraft sans key electronic components. The inability to conduct full tests of such assemblies early in a program is a source of technical and cost risk in spacecraft developments.

[b]SSR manufacturers must closely monitor market developments and dedicate resources to the ongoing process of evaluating new components to remain flexible in the face of supply changes.

Other factors, such as thermal degradation, are not well understood. Long-term storage of commercial EEE parts is also of some concern. Many PA personnel interviewed during the course of this study felt that factors associated with the long-term storage of commercial (plastic) components were not a serious issue for spacecraft, since parts are usually procured fresh, used quickly, and then delivered for launch. This perspective overlooks the possibility of unexpected storage, such as occurred on the Galileo and SWAS missions.[11] Additional research will evaluate these factors and provide the information needed for accurate assessments of risk.

IMPLICATIONS FOR FUTURE SPACECRAFT

Increasing reliability at the component level should create parallel improvements in the overall reliability of spacecraft. Smaller spacecraft offer additional reliability advantages, since reliability generally goes up as feature size diminishes. This trend is manifested in several ways:

- Higher levels of integration, while requiring more complex designs and an increased attention to thermal management, generally lead to increased reliability by reducing the number of subassemblies and the wiring needed to connect them. This translates into fewer points for failure and a reduced testing workload.

- Spacecraft are, increasingly, digital microelectronic assemblies. The increasing reliability of these components should translate into higher overall system reliability and the development of long-lived spacecraft.

- As spacecraft get smaller and lighter, structural loads are reduced, especially during launch and injection maneuvers. If electrical loads are also reduced, power systems have to do less work and less heat has to be rejected.

Components used in all spacecraft are also likely to be "smarter" and more failure-tolerant than past units. Fault-tolerant components and operating software will become more readily available for incorporation into spacecraft systems. Spacecraft automation developments are especially important in this regard.

SUMMARY

The civil, military, and commercial space sectors are focusing attention on the development of high-reliability space systems. Processes designed to identify

---

[11]The SWAS spacecraft was originally planned to be launched in June of 1995 but was delayed. SWAS is not expected to be launched until early 1999.

and apply high-reliability components, aided by insights gained from the reliability practices of other industries, promise continued reductions in the number and severity of spacecraft failures.

Designing for reliability has a corollary effect: Spacecraft tend to live well beyond original design points. Longer-lived spacecraft create a challenge in terms of operating budgets; however, longevity can be beneficial in terms of offering new approaches to conducting space research, observing the unexpected, having resources on-hand to view emergent phenomena, supporting other missions, having greater mission planning flexibility, and training.

The increasing use of commercial components is of concern in terms of the future reliability of space systems. High-quality commercial parts are more available than their space-rated alternatives and typically offer greater performance. At issue is the long-term reliability of electronic devices encapsulated in plastic instead of traditional ceramics. Recent experience has shown that these components can meet the rigors of spaceflight, but certain areas, principally radiation resistance, remain that must be addressed through ongoing research. It is likely that supplies of space-rated components, especially high-performance microelectronics, will dwindle. Commercial electronic parts will, therefore, be increasingly important to the performance and reliability of future space systems.

Parts and components that are more reliable should translate into significant improvement in overall system reliability. Additionally, by their nature, small spacecraft offer advantages in terms of reliability, since smaller, more integrated systems have historically demonstrated higher reliability. The increasing proportion of microelectronic systems onboard future spacecraft should also lead to improved reliability, as should decreases in structural loads. It is possible, therefore, to envision future spacecraft that achieve unprecedented levels of performance through the use of systems designed expressly for high reliability.

NEW APPROACHES TO SPACECRAFT DESIGN

INTRODUCTION

The design phase lies at the core of any complex system. In relation to spacecraft, it encompasses not only the development of actual flight hardware and software but also the preceding mission definition, the procedures for test and operation, and, ultimately, a strategy for synthesizing data from the mission into useful scientific knowledge. Design is, therefore, a comprehensive process, the outcome of which depends on cooperation among a team of experts under the guidance of seasoned management.

Engineers have long experimented with new ways of designing systems, but the pressures that have resulted from shrinking the cost and size of spacecraft have greatly accelerated natural process improvement. The design process is now expected to deliver less-expensive, more-capable spacecraft. Achieving this while improving performance and reliability presents significant challenges to design teams.

This appendix will first review how constrained budgets have influenced approaches to design. It will then describe what steps have been taken to develop new, lower-cost design processes. The implications of new design approaches will also be discussed.

DESIGNING WITHIN CONSTRAINED BUDGETS

The primary driver for improving the design process is cost. Design engineering is the largest element in the overall cost of building space systems. Typically, 60 percent of the budget for building a spacecraft is expended prior to fabrication. (Wong, 1992, p. 734.) As discussed in Appendix A, the small spacecraft NASA is currently building retain much of the complexity of their larger predecessors. Not surprisingly, nonrecurring costs remain typically 50 to 60 percent of the cost of a small spacecraft. (Bearden, 1996, p. 44.)

Because design is a major factor in the cost equation, managers must be especially careful to avoid growth in this area. One element of a "design-to-cost" strategy is to maintain a firm cutoff in the amount of engineering allowed for a given spacecraft. Designs are frozen early, and attention is shifted to the test phase. "Testing the hell out of the design" has always been an element of spacecraft engineering, but many small spacecraft rely on testing to an exceptional degree. Trading design costs for additional testing can help mitigate risk, as demonstrated in the recent Mars Pathfinder mission. Mars Pathfinder's design was high risk in that it was single-string and relied extensively on new design approaches. (Muirhead, 1996, pp. 7–9.) To improve the probability of success, the spacecraft was rigorously tested prior to launch.

Controlling design costs is mandatory in a small spacecraft program, but to reduce costs, the design effort must be reduced. For example, many small programs forgo the development of the engineering test units that have traditionally been used to work out design and system-level bugs prior to committing to actual flight hardware. In the past, these test units took various forms— structural models, protoflight units, proof test models, etc.—but today they have been replaced by less expensive analytical models.

Reductions in the design effort can work against other mission objectives. Shortening the design phase can, for example, limit the ability of a program to incorporate advanced technology. When all mission elements are considered, advanced designs are sometimes rejected—not because of fear that new components will fail, but because of the time it takes to integrate them. This is potentially limiting in that future small spacecraft are expected to deliver increasingly impressive performance. Less attention to design can also adversely affect reliability. Appendix B showed that design errors are the major source of failure in spacecraft systems. This suggests that the design phase should be the focal point for risk reduction and urges more, not less, attention to design.

Close monitoring of design costs and keeping the design phase as brief as practical are strategies that have helped bring down the cost of building spacecraft. To reduce costs further, however, new processes are needed that reduce the time required to perform the engineering function and speed the incorporation of new technology. To reduce risk, new processes also need to integrate (a) the knowledge gained from advanced failure-analysis efforts, (b) the results of improved test strategies, and (c) information related to high-reliability parts and components.

RETHINKING THE DESIGN PROCESS

Rethinking the design process means changing not only the drawing-board phase, when actual engineering drawings are prepared, but also the earlier mission planning phase, when critical trades and selections are made. The SMEX program, one of NASA's premier small spacecraft programs, recognizes that "the mission design, not just the spacecraft, must be optimized to reduce the workload and to shorten the development/integration/test activities." (Watzin, 1996b, p. 2.)

One approach to living with limited budgets while attempting to mitigate risk and increase performance is to attempt to spread design costs across a vertically integrated program, achieving some degree of cost recovery. NASA's Explorer program employs this approach. Each new Explorer spacecraft, regardless of size, builds on the heritage of the past; each new design introduces features with the next unit in mind. Purposely designing systems to scale up or down helps to minimize the time, and thus the cost, of maturing a design for use on a new mission. Errors in design can also be eliminated in future versions of a given design. This approach has paid dividends, as demonstrated by the history of performance and reliability of Explorer spacecraft.

Collaborative Approaches to Design

One of the most important improvements related to design is a greater degree of collaboration within design teams. Underlying this shift is a fundamental change in how engineers view spacecraft systems. A traditional definition of a spacecraft would be based on a hierarchical view of discrete systems that communicate through predefined interfaces. A current definition might view spacecraft systems as interrelated, dynamic, and reconfigurable.

The traditional view of spacecraft relied heavily on a work breakdown structure (WBS), a top-down strategy for approaching the design task. Each WBS element represented a discrete design element, and engineers were given budgets (both cost and technical, in terms of mass, power, etc.) that they were expected to stay within. Expert teams solved the design challenges of each element of the WBS in relative isolation. At predefined points in the schedule, the overall team, or subsets of the team at the system level, would gather to check on progress and share relevant information. There were some drawbacks to this approach:

- The WBS approach tended to focus on designing spacecraft systems; how it was to be operated and even how it was to be tested were often not considered until the design was nearly complete.

- Optimization was difficult and usually occurred only locally within the design.

- Segregating the design effort ignored the obvious interconnectedness of spacecraft systems and subsystems.

- A good deal of internal documentation and communication was required to define interfaces.

- Communication failures within the design team, formal or informal, often caused critical items to be overlooked, later necessitating expensive fixes and workarounds.

The presence of a WBS usually influenced the organization of the design team. Teams formed in a hierarchy found it inherently difficult to communicate, and they often acted competitively rather than cooperatively. Perhaps the biggest problem with the traditional model, however, was that elements of the design came together only periodically. This meant that managers could get an accurate picture of the overall progress only at prescribed review points defined by NASA's program management guidelines. (Casani et al., 1994, p. 230.)

Despite these limitations, the traditional approach worked and produced dramatic successes. Eventually, however, a more collaborative view of the design process began to take hold. System engineering became rooted in spacecraft design practices, an improvement that began to broaden the focus of the effort to include life-cycle considerations.[1] Establishing the system engineering function to integrate across the design and development processes was an important innovation, but it concentrated on improving the technical aspects of design and retained the inherent hierarchical organization of the effort.

When viewed collaboratively, the importance of subordinate elements shrinks in relation to the increasing importance of the whole system. This approach focuses on broad-scale goals; not just technical performance, but cost, risk, operability, manufacturability, and end use, are optimized within the design process.

Concurrent engineering and the corollary innovation, *integrated product teams* (IPTs), are manifestations of collaborative design. These techniques place less emphasis on hierarchical team organization and linear approaches to design. Formal design phases (mission concept to preliminary design to final design) are replaced by an iterative process in which designers, test engineers, operators, and mission planners communicate directly and form multidisciplinary teams. This approach is well suited to the small spacecraft environment, in which many variables must be optimized. Indeed, many of the small spacecraft programs in this study have experimented with or wholly adapted concurrent engineering practices and the use of IPTs.

---

[1]Most of the small satellite programs reviewed in this study treated system engineering as a discrete element of the design process with its own budget.

Computer-Based Design Environments

Use of computer-based tools has expanded rapidly, helping to control design costs and reducing the need for test models. The majority of small spacecraft builders now use advanced design tools, such as the Computer-Aided Three-Dimensional Interactive Approach (CATIA) platform Boeing used to design the 777 aircraft. Although this capability is expensive, it is cost-effective in terms of reducing design time. Stand-alone design tools like CATIA can be limited, however, in their ability to interact with modeling and simulation (M&S) systems.

Advanced M&S systems began to reach a high state of fidelity in the early 90s and are a natural evolution of independent computer-based design tools. Collaborative approaches to design have proven to be well matched to advances in the M&S field. JPL and the then–Martin Marietta Corporation were both innovative in the creation of spacecraft design environments with extensive M&S capabilities. Martin Marietta's Spacecraft Technology Center (STC) in Denver promoted an intensive team environment in which aspects of mission design, spacecraft design, manufacturing, and operation could be quickly evaluated. The STC also made use of the Internet to exchange information and connect designers in remote locations into interactive design sessions. The initial STC was reconfigured with more advanced equipment and is now operating as the second-generation STC II.

At JPL, two related elements were created: the Project Design Center (PDC) and the Flight System Testbed (FST). These facilities were constructed with a stated goal of "recrafting" the engineering design process. JPL began by reevaluating all internal processes, breaking them down into four areas: project planning and implementation; mission and system design, fabrication, assembly, and test; and validation, integration, and operation. (Smith, 1996, p. 4.)

The PDC, shown in Figure E.1, is dedicated to what has traditionally been called mission design, the refinement of a science concept into a viable engineering design. To encourage team involvement, the PDC consists of one large room with peripheral support areas. An assortment of computers throughout the area allows engineers to run a suite of software tools and models, many of which are commercial tools, and to project results on large screens. As fitting the mission-design role of the PDC, these tools are selected to allow the team to perform trajectory studies, assemble power and mass budgets, generate solid models of the spacecraft, and estimate resulting costs. JPL relies on an expert technical body called the Advanced Projects Design Team, or Team X, made up of senior technical personnel, to assist in the initial design of a mission. Team X makes extensive use of the PDC to ensure that such issues as cost and operability are included in the overall mission design.

SOURCE: JPL.

Figure E.1—Team X in the Project Design Center

The FST, shown in Figure E.2, is a functional, system-level simulation of a proposed spacecraft. It contains computer-based analogs for each of the primary systems on the spacecraft, including the instruments. Simulations of the ground-control systems and the data-communication networks are also provided. The goal of the FST is to deliver a ready-to-build design that can be produced at reduced cost and schedule risk. This form of advanced simulation also allows new technology to be evaluated in modes similar to what will be experienced on the spacecraft.

To the extent that designs exist only inside of a computer (the term "silicon spacecraft" is often used), a virtual design environment is possible, one in which team members need not be physically colocated. Lockheed Martin Missiles and Space Company's Palo Alto Research Laboratory is pioneering this type of capability under the support of DARPA. What is called the Simulation Based Design (SBD) Laboratory is actually a geographically disperse collection of teams collaborating on the design of a product through Internet connections.

Members of the design team communicate electronically, making individual contributions to the overall design. High-performance computers render designs, perform structural analysis, calculate performance against objectives, and coordinate and update design information.

One of the principal challenges of such approaches has been the difficulty of linking together advanced design tools and simulation models into a single, in

SOURCE: JPL.

Figure E.2—Design Team in JPL's Flight Systems Testbed

teractive environment. Such linkage requires creating interface standards that allow disparate models to exchange information and operate interactively. Standards are emerging, such as the Common Object Request Broker Architecture (CORBA). Extensive use is also being made of current Internet standards, such as the HyperText Markup Language (HTML) and Virtual Reality Modeling Language (VRML), and interfaces that are familiar to a broad user community, such as WWW browsers like Netscape.

In the future, the SBD environment will likely be linked to data archives containing a common set of information on the parts and components used to build spacecraft. The result will be an enclosed design process where a complete spacecraft team can quickly close on a desired design solution and enter the fabrication and test phase with a high degree of confidence.

Many organizations are getting involved in the process of creating new design environments. NASA GSFC, for example, has recently established the Integrated Mission Design Center (IMDC). DARPA and the National Institute of

Standards and Technology are also funding extensive studies in manufacturing that are tied to virtual design environments.

Remaining Challenges

Advanced design processes are an important development in terms of helping builders deliver less-expensive, more-capable, and more-reliable spacecraft. Yet, these capabilities are costly to develop, and their availability could be a factor constraining use. Cultural factors must also be addressed to achieve widespread acceptance of the computer-based approaches.

The development of new design capabilities requires a level of investment that is likely to be beyond the means of many commercial developers of small spacecraft. JPL's management realized that smaller missions could not afford to contribute to significant improvements in the infrastructure needed to construct the PDC and FST. (Sander, 1997, p. 4.) The FST and PDC are, therefore, available at modest cost to in-house design teams. Pricing and prioritization policies for use by customers outside of JPL, however, have not yet been established.

Most small spacecraft programs have rapid development schedules and commensurately short design timelines. Facility priority is usually given to in-house projects; attractive pricing might, therefore, be of little use because of scheduling problems. A related example, access to test facilities, illustrates this point. A small spacecraft design team cannot usually accept uncertainty in the availability of a test facility and will often pay a premium for ensured access. To the extent that advanced design environments represent national assets, pricing and availability policies will need to be established.

The creation of new design facilities also requires the resources to experiment with alternative structures. The first incarnation of the PDC, for example, was found to be uncomfortable for design teams. Acoustics and lighting were poor, and the physical layout was not conducive to team operations. Subsequently, the PDC was moved twice to reach an arrangement that worked.

Finally, new approaches to design can encounter cultural barriers. When JPL's facilities were first opened in 1994, engineers used the new virtual environments as extensions of traditional practices, and their full potential was not realized (Smith, 1997, p. 3). Managers, too, can resist change. JPL project managers, traditionally able to select their own design approaches, faced standardization and the subsequent loss of autonomy (Smith, 1997, p. 4). Also, most programs still rely on mission-specific test beds for evaluating designs.[2] Design teams will likely require that new design environments demonstrate a clear advantage over mission-specific test beds before being willing to completely adopt them.

IMPLICATIONS FOR FUTURE SPACECRAFT

Virtual design environments are the state of the art, and spacecraft builders are only now starting to use them. As they gain acceptance, industry analysts predict significant cuts in design times and cost, while improving final component performance and reliability.

The ability of a virtual environment to help the engineer visualize the effects of design changes is the real advantage of working in a simulated environment. Feedback is rapid, and other team members are available to resolve problems and make the required trades. The goal of LMC's SBD system, for example, is to "reduce satellite design processes from months to days." (Graves et al., 1997, p. 7.) Such dramatic reductions are often difficult to achieve in practice, but there is ample reason to believe that large reductions in design times are possible. Intel, while achieving the quality and reliability targets reported in Appendix D, reduced the average component design time from 80 weeks in 1986 to 23 weeks in 1995. (Intel, 1996, pp. 1–8.)

The existence of computer-based design environments also offers an opportunity to integrate the factors described in Appendixes B and C. New test approaches, insights into sources of failure, and knowledge gained from research into high-reliability systems can be brought together in a central location that is coincident with the design effort.

SUMMARY

The design of space systems is a comprehensive process that is being reengineered to deliver less-expensive, more-capable spacecraft that perform better and offer greater reliability. In regard to space systems, cost is the primary driver for changing the design process, since the design phase is typically the most expensive cost element in spacecraft TMC.

---

[2]Test beds are a synthesis of computer models and physical elements that simulate the operation of the spacecraft and its ground control network. As the development of the spacecraft continues, simulated systems are replaced by actual flight equipment, so the test bed serves as both a design and test tool. Test beds are often built specifically to meet the requirements of a mission. Advanced design environments seek to provide many of the capabilities of the mission-specific test beds, reducing or eliminating the need for them.

Builders of small spacecraft are especially pressed to minimize the length, and thus the cost, of the design phase. Some of the methods used to control design cost are

- Capping the design effort (design-to-cost) and focusing on testing
- Forgoing the use of engineering test units
- Reducing new technology in the design.

These methods can work against other goals, such as reducing design-related failures and increasing the performance of spacecraft systems. New design approaches seek to improve the cost and technical effectiveness of the design process.

One of the most important improvements has been a greater degree of collaboration within design teams. The traditional hierarchical design process, built around the work breakdown structure, has been largely replaced by a collaborative process. RAND found that most of the small spacecraft programs in this study have reflected this shift by experimenting with or wholly adapting concurrent engineering practices and the use of integrated product teams.

Design process improvement has been paralleled by gains in the performance of modeling and simulation tools. Initial developments in this area have centered around the creation of design centers in which engineers are immersed in a team environment, surrounded by the latest computer-based tools. JPL's Project Design Center and Flight System Testbed are representative of such developments.

A natural extension of such centers is to connect geographically disperse teams via the Internet. Such "virtual" design environments connect teams via high-speed, fiber-optic links. Engineers can quickly analyze aspects of the emerging design by accessing local or remote tools, make changes, and communicate them to other team members.

The emergence of a collaborative design process, supported by computer-based environments containing advanced modeling and simulation tools, is an important development in terms of reducing the cost and risk associated with space systems.

SMALL SPACECRAFT WORLD WIDE WEB LINKS

SPACE MISSION INFORMATION

Mission Compilations

Earth Observing System Home Page	eos.nasa.gov/
Discovery Program Homepage	discovery.larc.nasa.gov/discovery/home.html
Earth System Science Pathfinder Project	essp.gsfc.nasa.gov/
MIDEX Program Office	www710.gsfc.nasa.gov/Projects/MIDEX/xhome.html
New Millennium Program (NMP)	nmp.jpl.nasa.gov/
Small Spacecraft Technology Initiative (SSTI) Program	www.crsp.ssc.nasa.gov/ssti/welcome.htm
Dan's Unmanned Spacecraft Links	sulu.lerc.nasa.gov/dglover/craft.html
Land Observation Satellites	geo.arc.nasa.gov/esdstaff/landsat/wes.html
Mike's Spacecraft Library	leonardo.jpl.nasa.gov/msl/
Satellites (CY 1996)	www.aero.org/activities/satellites.html
Space Mission Acronym List and Hyperlink Guide	ranier.oact.hq.nasa.gov/Sensors_page/ MissionLinks.html#NMP
Spacecraft Data Book - OSS	www.hq.nasa.gov/office/oss/enterprise/index.html
ACE Home Page	www.gsfc.nasa.gov/ace/ace.html
ALEXIS Project	nis-www.lanl.gov/nis-projects/alexis/
AXAF Home Page	hea-www.harvard.edu/asc/axaf-welcome.html
Cassini: Voyage to Saturn	www.jpl.nasa.gov/cassini/
CATSAT Satellite	burst.unh.edu/CATSAT/catsat.html
Clementine I	nssdc.gsfc.nasa.gov/planetary/ lunar/clementine1.html
Cluster Project	www-istp.gsfc.nasa.gov/ISTP/cluster_project.html
COBE Home Page	www.gsfc.nasa.gov/astro/cobe/cobe_home.html
CRRES Mission	tide1.space.swri.edu/crres.html
Defense Support Program	www.laafb.af.mil/SMC/PA/Fact_Sheets/dsp_fs.htm
DMSP—Defense Meteorological Satellite Program	www.aero.org/dmsp/
DSCS Homepage	www.laafb.af.mil/SMC/MC/DSCS/
Earth Orbiter 1 (EO-1)	www511.gsfc.nasa.gov/eo-1/
Far InfraRed and Submillimetre Telescope (FIRST)	astro.estec.esa.nl/SA-general/Projects/First/first.html
FUSE Home Page	fuse.pha.jhu.edu/
FUSE Homepage	profuse.pha.jhu.edu/
Geotail	www-spof.gsfc.nasa.gov/istp/geotail/index.html
GGS Project (WIND+POLAR)	www-istp.gsfc.nasa.gov/ISTP/ggs_project.html

GLAST: The Gamma Ray Large Area Space Telescope	www-glast.stanford.edu/
Global Geopsace Science (GGS) Program	ggsfot.gsfc.nasa.gov/
GPS/MET HOME PAGE	pocc.gpsmet.ucar.edu/
Gravity Probe—B (GP-B) Home Page	stugyro.stanford.edu/RELATIVITY/GPB/GPB.html
GSFC Space Science Mission Descriptions	marconi.gsfc.nasa.gov/gsfc/spacesci/sentinel/spacesen.htm
HESI—High-Energy Solar Imager	umbra.nascom.nasa.gov/solar_connections/HESI.html
High Throughout X-Ray Spectroscopy (HTXS)	htxs.gsfc.nasa.gov/
HIRDLS Home Page	eos.acd.ucar.edu/hirdls/home.html
HTXS—Science	htxs.gsfc.nasa.gov/docs/xray/htxs/science.html
INFLATABLE ANTENNA EXPERIMENT	www.jpl.nasa.gov/iae/
International Space Station	station.nasa.gov/
ISTP Program	www-istp.gsfc.nasa.gov/
JPL Scatterometer	winds.jpl.nasa.gov
Landsat Program	geo.arc.nasa.gov/esdstaff/landsat/landsat.html
LP—Lunar Prospector @ LMC	juggler.lmsc.lockheed.com/lunar/
LP—Lunar Prospector @ Ames Research Center	pyroeis.arc.nasa.gov/lunar_prospector/home.html
Magellan Mission to Venus (JPL)	www.jpl.nasa.gov/magellan/
Mars Pathfinder	mpfwww.jpl.nasa.gov/
Meteorological Satellite Page	www.met.nps.navy.mil/bob/sat/met_sat.html
MGS—Mars Global Surveyor Project	mgs-www.jpl.nasa.gov/
MIDEX Home Page	midex.gsfc.nasa.gov/
MSX Home Page	sd-www.jhuapl.edu/MSX/MSX_Overview.html
MSX Satellite	msx.nrl.navy.mil/
MTPE EOS AM-1/2	fpd-b8-0001.gsfc.nasa.gov/421/421proj.htm
MTPE EOS Chem-1	fpd-b8-0001.gsfc.nasa.gov/424/424proj.htm
MTPE EOS PM-1	fpd-b8-0001.gsfc.nasa.gov/422/422proj.htm
NEAR Mission	hurlbut.jhuapl.edu:80/NEAR/
NMP Deep Space 1	nmp.jpl.nasa.gov/Missions/DS1/
NMP Deep Space 2	nmp.jpl.nasa.gov/Missions/DS2/
Pluto Express Mission	www.jpl.nasa.gov/pluto/
SAMPEX	lepsam.gsfc.nasa.gov/www/sampex.html
SH/MPS On-Line Home Page	www.microprose.com/
SIM—Space Interferometry Mission	huey.jpl.nasa.gov:80/sim/
SIRTF Home Page	sirtf.jpl.nasa.gov/sirtf/home.html
SMEX Program Office	sunland.gsfc.nasa.gov/smex/smexhomepage.html
SNOE Spacecraft	miranda.colorado.edu/snoe/
SOFIA Home Page	sofia.arc.nasa.gov/
SOHO Mission	sohowww.nascom.nasa.gov/
SPARTAN HOME PAGE	spartans.gsfc.nasa.gov/
SSTI/Lewis	crsphome.ssc.nasa.gov/ssti/welcome.htm
STARDUST Mission	pdcsrva.jpl.nasa.gov/stardust/home.html
TERRIERS Spacecraft	net.bu.edu/terriers/terriers.html
TIMED Spacecraft	umbra.nascom.nasa.gov/solar_connections/TIMED.html
TiPS: Tether Physics and Survivability Satellite Experiment	hyperspace.nrl.navy.mil/TiPS/
TOPEX/Poseidon	podaac-www.jpl.nasa.gov/tecd/pop.htm
TOPHAT	cobi.gsfc.nasa.gov/msam-tophat.html
TRACE Spacecraft	www.space.lockheed.com/TRACE/welcome.html

Tropical Rainfall Measuring Mission (TRMM)	fpd-b8-0001.gsfc.nasa.gov/490/490home.htm
TRMM Office Home Page	trmm.gsfc.nasa.gov/trmm_office/index.html
UARS	uarsfot08.gsfc.nasa.gov/UARS_HP.html
UARS PROJECT DEFINITION	daac.gsfc.nasa.gov/CAMPAIGN_DOCS/ UARS_project.html
Wake Shield Facility	www.svec.uh.edu/wsf.html
WIRE	sunland.gsfc.nasa.gov/smex/wire/index.html
XTE Home Page	heasarc.gsfc.nasa.gov/docs/xte/xte_1st.html

NASA ORGANIZATIONS

Ames Research Center

ARC Center for Mars Exploration	cmex-www.arc.nasa.gov/
NASA Ames Research Center	www.arc.nasa.gov/

Goddard Space Flight Center

NASA/Goddard Projects and Organizations	www.nasa.gov/GSFC_orgpage.html
Code 400—Flight Projects Directorate	fpd-b8-0001.gsfc.nasa.gov/400/400home.htm
Code 500—Mission Operations & DS Directorate	joy.gsfc.nasa.gov/
Code 600—Space Sciences Directorate	ssdoo.gsfc.nasa.gov/c600/c600.html
Code 700—Engineering Directorate	server701.gsfc.nasa.gov/homepage.html
Code 704—Systems Engineering Welcome Page	www710.gsfc.nasa.gov:80/704/
GSFC—Mission To Planet Earth	mtpe.gsfc.nasa.gov/
GSFC Library top homepage	www-library.gsfc.nasa.gov/
Autonomous Spacecraft Operations—GSFC	talos.stel.com/
Explorer Program Office	www710.gsfc.nasa.gov/Projects/EXPLORER/
GGS Mission Operations	ggsfot.gsfc.nasa.gov/
Goddard Space Flight Center Home Page	pao.gsfc.nasa.gov/gsfc.html
GSFC— EEE Parts Link	arioch.gsfc.nasa.gov/eee_links/vol_01/no_03/eee1-03.html
GSFC Spacecraft Reliability Page	arioch.gsfc.nasa.gov/302/relhp.htm
GTRS—Goddard Technical Report Server	www-library.gsfc.nasa.gov/Gtrs/Gtrs-home.html
Mission to Planet Earth	mercury.hq.nasa.gov/office/mtpe/
NASA Mission Design Process Page	www710.gsfc.nasa.gov/704/grnbook/grnbook.html
NASA/GSFC Assurance Technologies Division Home Page	arioch.gsfc.nasa.gov/310/310.html
Small Satellite Technology Page	www.gsfc.nasa.gov/gsfc/small_sat/small_sat.html
System Reliability Office	arioch.gsfc.nasa.gov/302/srsohp4.htm

Headquarters

Office of the Administrator

The NASA Homepage	www.nasa.gov/index.html
NASA Headquarters Home Page	venus.hq.nasa.gov/
NASA Materials and Processes Homepage	nasa-mp.jpl.nasa.gov/jpl-mp/homepage.htm

Office of the Chief Financial Officer

Code B Home Page	venus.hq.nasa.gov/office/codeb/
Integrated Financial Management Project	booster.nasa.gov:443/
NASA FY 1997 Budget	www.hq.nasa.gov/office/codeb/budget/

Office of Headquarters Operations

NASA Headquarters Library Home Page	www.hq.nasa.gov/office/hqlibrary/
NASA Technical Report Server (NTRS), v 2.0	techreports.larc.nasa.gov/cgi-bin/NTRS

Office of Procurement

NASA Office of Procurement Home Page	www.hq.nasa.gov/office/procurement/welcome.html
NASA HQ Procurement—NAIS	procurement.nasa.gov/
NASA Acquisition Forecast	procure.msfc.nasa.gov/forecast/index.html

Office of Management Systems and Facilities

Management Instructions and Handbooks	lincoln.gsfc.nasa.gov:80/rnd/Welcome.html

Office of Space Flight

OSF Home Page	www.osf.hq.nasa.gov/
Mixed Fleet Manifest Home Page	www.osf.hq.nasa.gov/manifest/
Space Launch List	www.osf.hq.nasa.gov/1997/launch97.html

Office of Public Affairs

NASA Photo Gallery	WWW.HQ.NASA.GOV/office/pao/Library/photo.html
NASA Public Affairs	www.gsfc.nasa.gov/hqpao/hqpao_home.html
NASA Image eXchange (NIX)	nix.nasa.gov/

Office of Safety and Mission Assurance

OSMA Home Page	www.hq.nasa.gov/office/codeq/

Office of Space Science

OSS Advanced Technology & Mission Studies Division	www.hq.nasa.gov/office/oss/osstech/techhome.htm
Advanced Technology and Mission Studies	venus.hq.nasa.gov/office/oss/osstech/techhome.htm
Astrophysics Main Page	venus.hq.nasa.gov/office/astrophysics/
Discovery Program Office	mercury.hq.nasa.gov/office/discovery/
OSS Homepage	www.hq.nasa.gov/office/oss/
Solar Connections Program	umbra.nascom.nasa.gov/solar_connections.html
Solar System Exploration Theme Page	venus.hq.nasa.gov/office/oss/solar_system/
Space Physics Homepage	umbra.nascom.nasa.gov/spd/spd.html

Office of Space and Advanced Technology

OSAT Advanced Concepts Office	www.hq.nasa.gov/office/acrp/oac.html
OSAT Spacecraft Systems Division	ranier.oact.hq.nasa.gov/SCRS_Page/SCHP.html

Office of Mission to Planet Earth

| Mission to Planet Earth | www.hq.nasa.gov/office/mtpe/ |
| Mission to Planet Earth Strategic Enterprise Plan 1996-2002 | www.hq.nasa.gov/office/mtpe/stratplan/stratplan.html |

Office of Policy and Plans

| OPP Home Page | www.hq.nasa.gov/office/codez/ |
| NASA Policy Page | www.hq.nasa.gov/office/codez/policy.html |

Jet Propulsion Laboratory

Jet Propulsion Laboratory Home Page	www.jpl.nasa.gov/
Technical Reports Server (JPLTRS)	techreports.jpl.nasa.gov/index.html
Center for Space Microelectronics Technology	mishkin.jpl.nasa.gov/newcsmt.html
Flight System Testbed	fst.jpl.nasa.gov/
Comet Hale-Bopp Home Page (JPL)	www.jpl.nasa.gov/comet/index.html
Flight Performance Assessment	techinfo.jpl.nasa.gov/www/fpa/fpa.html
Flight Systems Project Office	atlas2.jpl.nasa.gov/
Project Design Center	pdc.jpl.nasa.gov/
Space and Earth Science Directorate	www.jpl.nasa.gov/sespd/
New Millennium Program	nmp.jpl.nasa.gov/
Second Generation MicroSpacecraft	edms.jpl.nasa.gov/msc/

Johnson Space Center

| JSC Home Page | www.jsc.nasa.gov/ |
| NASA—JSC Digital Image Collection Home | images.jsc.nasa.gov/html/home.htm |

Kennedy Space Center

| NASA Expendable Launch Vehicle Info | www.ksc.nasa.gov/elv/elvpage.htm |

Marshall Space Flight Center

| MSFC Safety and Mission Assurance (S&MA) | msfcsma3.msfc.nasa.gov:8001/s_ma.html |

MILITARY SPACE ORGANIZATIONS

Advanced Research Projects Agency	www.arpa.mil/
AFSMC Chief Engineer's Page	sdf.laafb.af.mil/~gowerj/chief_engineer.html
Air Force Link—Official web site of the United States Air Force	www.af.mil/
Air Force Phillips Laboratory	www.plk.af.mil/
Air Force Rome Laboratory Home Page	www.rl.af.mil/
Air Force Space Command	www.spacecom.af.mil/hqafspc/
Army Missile Defense and Space Technology Center	www.ssdc.army.mil/SSDC/MDSTC.html
Army Space and Strategic Defense Command	www.ssdc.army.mil/
Army Space Command	armyspace.com/

BMDOLINK	www.acq.osd.mil/bmdo/bmdolink/html/bmdolink.html
College of Air and Space Doctrine—CADRE	www.cdsar.af.mil/
DARPA	www.arpa.mil/
Defense Information Systems Agency (DISA)	www.disa.mil/disahome.html
DoctrineLINK—Military Analysis Network	www.fas.org/man/doctrine.htm
DUSD Space	www.acq.osd.mil/space/
Future Warfare @ DTIC	www.dtic.mil/doctrine/jv2010/
Information Warfare	www.infowar.com/
Los Alamos—NIS Division	nis-www.lanl.gov/
MILSATCOM Homepage	www.laafb.af.mil/SMC/MC/
NATIONAL RECONNAISSANCE OFFICE	www.nro.odci.gov/
Naval Center For Space Technology	www.nrl.navy.mil/nrl/direct/code.8000.html
Naval Space Command Home Page	www.navy.mil/homepages/navspacecom/
NAVSTAR GPS HOMEPAGE	www.laafb.af.mil/SMC/CZ/homepage/
NORAD Space Objects Data Base	oigsysop.atsc.allied.com/
North American Aerospace Defense Command	www.spacecom.af.mil/norad/
Office of Naval Research	www.onr.navy.mil/
OSD Acquisition and Technology	www.acq.osd.mil/
OSD Open Systems Task Force	www.acq.osd.mil/osjtf/
Space Testbed—SMC/TE	www.te.plk.af.mil/
SPAWAR's Home Page	www.spawar.navy.mil/
The Naval Research Laboratory	www.nrl.navy.mil/
US Navy SPAWAR	www.nosc.mil/spawar/welcome.page
US Space Command Home Page	www.spacecom.af.mil/usspace/
Wright-Patterson Air Force Base Bulletin Board	www.wpafb.af.mil/

SPACECRAFT MANUFACTURERS

Hughes

Hughes Communications WebStation	www.hcisat.com/index.html
Hughes Space and Communications	www.hughespace.com/
Hughes STX Corporation	www.stx.com/
Hughes-STX Software Engineering Lab	info.stx.com/

Lockheed Martin

Lockheed Martin—Astronautics	www.lmco.com/Astro/home.html
Lockheed Martin Corporation	www.lockheed.com/
Lockheed Martin Missiles and Space (Sunnyvale)	www.lmsc.lockheed.com/
Lockheed-Martin Solar and Astrophysics	www.space.lockheed.com/9130.html

TRW

TRW Inc.	www.trw.com/
TRW Products & Services: Space/Satellites	www.trw.com/prod_serv/space/index.html
TRW Smallsat Main Page	marvin.sedd.trw.com:1234/
Aerospace Corporation	www.aero.org/index.html

AGI—Satellite Tool Kit	www.stk.com/
AlliedSignal Aerospace	www.alliedsignal.com/aerospace/index.html
Ball Aerospace & Technologies Corp.	www.ball.com/bhome/hspacebu.html
Boeing Company	www.boeing.com/
COMSAT Corporation	www.comsat.com/corp/home/Comsat.html
Corporate Satellite Links	www.pxi.com/public/links/links_orgs.html
CTA, Inc.	www.cta.com/
ITHACO Inc.	www.newspace.com/Industry/Ithaco/home.html
ITRI Satellite Communications Technology Panel	itri.loyola.edu/satcom/toc.htm
Jackson and Tull	www.jnt.com/
Litton Amecom	www.amecom.com/
Malin Space Science Systems	barsoom.msss.com/
Microcosm, Inc.	www.sblink.com/microcosm
Motorola Space	www.mot.com/
New Space Network	www.newspace.com/home-hi.html
Orbital Sciences Corporation's (OSC's) Home Page	www.orbital.com/
SAIC Space Physics Group	www.nw.saic.com/spacegrp.html
Satellite Tool Kit (STK) by Analytical Graphics	www.stk.com/
Schaeffer Magnetic, Inc.	www.schaeffer.com/
Software Technology, Inc.	www.mlb.sticomet.com/sti_hist2.htm
SoHar	fermi.sohar.com/spacesys.html
Spacehab	www.spacehab.com/corporate/
SPACEHAB HOME PAGE	208.196.158.205/
Spectrum Astro	www.spectrumastro.com/
Swales & Associates Inc.	www.swales.com/

UNIVERSITY SPACECRAFT ORGANIZATIONS

Boston University Center for Remote Sensing	crs-www.bu.edu/
Boston University Center for Space Physics	bu-ast.bu.edu/csp.html
Center for Space Telemetering at New Mexico State University	aloha.nmsu.edu:80/telemetry/
CU Structural Dynamics and Control Lab	sdcl-www.Colorado.EDU/
High Energy Astrophysics—Harvard	hea-www.harvard.edu/
ISU—International Space University	isu.isunet.edu/
JHU Applied Physics Lab	sd-www.jhuapl.edu/
MIT Center for Space Research	space.mit.edu/
New Mexico State—Physical Science Lab	www.psl.nmsu.edu/Welcome.html
Smithsonian Astrophysical Observatory	cfa-www.harvard.edu/sao-home.html
Space Telescope Science Institute Home Page	www.stsci.edu/top.html
Stanford International Mars Program	www-leland.stanford.edu/group/Mars/
UK Surrey: Small Satellites	www.ee.surrey.ac.uk/EE/CSER/UOSAT/SSHP/sshp.html
Univ of Maryland Space Systems Laboratory	www.ssl.umd.edu/
Universities Space Research Association (USRA)	www.usra.edu/

University of Colorado Structural Dynamics and Control Lab	sdcl-www.Colorado.EDU/
University of New Hampshire Small Satellite Lab Home Page	burst.unh.edu/ssl.html
US Air Force Academy Department of Astronautics	www.usafa.af.mil/dfas/dfas_hom.html
USU Department of Mechanical and Aerospace Engineering	www.engineering.usu.edu/Departments/mae/
Weber State Center for AeroSpace Technology	137.190.32.131/

INTERNATIONAL SPACE ORGANIZATIONS

AMSAT-NA	www.amsat.org/
AspireSpace—The British Amateur Space Rocket / Programme	www.gbnet.net/orgs/aspire/
Centre National Etudes Spatiales	www.cnes.fr/
European Space Agency	www.esrin.esa.it/
International Association for the Astronomical Arts	www.novaspace.com/IAAA/IAAA.shtml
ISAS Japan: Center for Advanced Spacecraft Technology	www.isas.ac.jp/index-e3.html
Kashima Space Research Center	clipper.crl.go.jp/
National Space Development Agency of Japan	www.nasda.go.jp/welcome_e.html
Russian Space Science Institute	www.rssi.ru/
Satellite Journal International	www.nmia.com/~roberts/sji/sj.html
Satellites International Home Page	www.sil.com/home.htm
UK Surrey: Small Satellites	www.ee.surrey.ac.uk/EE/CSER/UOSAT/SSHP/sshp.html
United Nations—Office for Outer Space Affairs	www.un.or.at/OOSA_Kiosk/
United Nations Office for Outer Space Affairs	www.un.or.at/OOSA_Kiosk/index.html
World Satellite Network	www.worldsat.com/

SPACECRAFT RELIABILITY

Advanced Test Engineering, Inc.	www.besttest.com/
ASTM Web Site	www.astm.org/
CALCE Electronic Packaging Research Center	spezia.eng.umd.edu/
Components Test Equipment Design and Manufacture	floyd.os.kcp.com/home/catalog/comrelab.html
COSMIC—NASA's Software Technology Transfer Center	cognac.cosmic.uga.edu/
DCMC Main Page	www.dcmc.dcrb.dla.mil/
DEF CON	www.defcon.org/
Defense Logistics Agency—Materiel Management	www.supply.dla.mil/
Defense Special Weapons Agency— ERRIC Radiation Data Base	erric.dasiac.com/
DoD Standards and Specifications	www.dodssp.daps.mil/
Electronic Industries Association	www.eia.org/
Electronic Visualization Laboratory	www.evl.uic.edu/EVL/index.html

Electronics Cooling Magazine On-Line	www.electronics-cooling.com/
Electronics Industry Environmental Roadmap: Table of Contents	www.mcc.com/env/roadmap/roadmap.toc.html
Electronics Quality Reliability Center	www.sandia.gov/eqrc/eqrc.html
FAA Center for Aviation Systems Reliability	www.cnde.iastate.edu/faa.html
Government-Industry Data Exchange Program (GIDEP) @ US Navy	www.gidep.corona.navy.mil/
GSFC—Radiation Effects & Analysis Home Page	flick.gsfc.nasa.gov/radhome.htm
GSFC Electronic Packaging & Processes Branch	arioch.gsfc.nasa.gov/312/312.htm
GSFC Office of Flight Assurance	arioch.gsfc.nasa.gov/
GSFC Preferred Parts List	arioch.gsfc.nasa.gov/ppl/ppl.htm
Harris Semiconductor—Reliability Page	rel.semi.harris.com/
IEEC Home Page	www.ieec.binghamton.edu/ieec/
IEEE Standards Home Page	standards.ieee.org/index.html
IEEE/CPMT Society Home Page	www.cpmt.org/
Institute of Environmental Sciences	onweb.com/cow/mall/institute.html
Intel's Quality System	developer.intel.com/design/news/quality.htm
International Standards Organization—ISO	www.iso.ch/welcome.html
ISO 14000 InfoCenter	www.iso14000.com/
MIDAS Homepage	www.lsi.edu/midas/midas.html
MTIAC Current Awareness Bulletin	mtiac.hq.iitri.com/mtiac/cab/
NASA EEE Part Information Management Systems (EPIMS)	epims.gsfc.nasa.gov/
NASA EEE Parts Program	nppp.jpl.nasa.gov/
NASA Lesson Learned Data Base	www710.gsfc.nasa.gov/704/lesslrnd/llitems/llindex.html
NASA Lessons Learned Home Page	envnet.gsfc.nasa.gov/ll/llhomepage.html
NASA R&M Steering Committee Home Page	www.hq.nasa.gov/office/codeq/rmhome23.htm
NASA Reliability Program	www.hq.nasa.gov/office/codeq/prctls23.htm
NCMA Homepage	www.ncmahq.org/
Network Reliability and Interoperability Council	www.fcc.gov/oet/nric/
QEMA, Quality, Engineering & Manufacturing Association	www.tqm.com/
Quality Online Home Page	qualitymag.com/
Radiation Effects & Analysis Home Page	flick.gsfc.nasa.gov/radhome.htm
Reliability Analysis Center—Rome AFB	rome.iitri.com/rac/
RELIABILITY magazine	www.reliability-magazine.com/index.phtml
Reliability Society Standards Development	stdsbbs.ieee.org/groups/reliability/index.html
Rome Lab—Electromagnetics & Reliability Home Page	erd.rl.af.mil/
SATC Home Page	satc.gsfc.nasa.gov/homepage.html
SATWG/SIIG home page	WWW.SATWG-SIIG.com/
SBD Homepage	sbdhost.parl.com/

Semiconductor Industry Associations website	www.semichips.org/
SEMI's Home Page	www.semi.org/
Space Environment Effects Branch	satori2.lerc.nasa.gov/
Space Environment: An Overview	satori2.lerc.nasa.gov/DOC/seeov/seeov.html
Space Testbed—SMC/TE	www.te.plk.af.mil/
Special Environments and Military	www.intel.com/design/specenvn/
TIs CARMA: Reliability & Maintainability Applications	www.ti.com/carma
University of Maryland Reliability Center	www.enre.eng.umd.edu/
US Army Materials Systems Analysis Activity	amsaa-www.arl.mil/rad/pofpage.htm

SPACE INFORMATION SOURCES

Aerospace Cost Estimation Programs	www.jsc.nasa.gov/bu2/ELV_US.html
Aerospace Navigator	www.ultranet.com/~adjm/aero/aeronav.html
AIAA Home Page	www.aiaa.org/
AIAA Publications and Databases	www.aiaa.org/publications/index.html
Air University Home Page	www.au.af.mil/
American Astronomical Society Home Page	www.aas.org/
American Society for Quality Control	www.asqc.org/
ANSI"—National Standards Systems Network	www.nssn.org/
Astronomical Society of the Pacific	www.aspsky.org/
Brookings—Final Report	www.in-search-of.com/frames/nasa_brookings/files_bell/report_pt1.html
CIESIN Home Page	www.ciesin.org/
CONSOLIDATED SPACE OPERATIONS CONTRACT	www.jsc.nasa.gov/somo/lib.html
Consultative Committee for Space Data Systems Home Page	www.nasa.gov/ccsds/ccsds_home.html
Defense Daily	www.defensedaily.com/
DoD Costing References Start Page	www.dtic.mil/dodim/costweb.html
DSN Home Page	deepspace.jpl.nasa.gov/dsn/
Earth Satellite Ephemeris Service	www.chara.gsu.edu/sat.html
EnviroNET: The Space Environment Information Service	envnet.gsfc.nasa.gov/
FLORIDA TODAY Space Online	www.flatoday.com/
GDP Deflator	www.jsc.nasa.gov/bu2/Inflation.html
GTRS—Goddard Technical Report Server	www-library.gsfc.nasa.gov/Gtrs/Gtrs-home.html
Guide to NASA Online Resources	nic.nasa.gov/nic/guide/
Hearings of the Committee on Commerce, Science, and Transportation	www.senate.gov/~commerce/hearings/hearings.htm
Inflation Calculator	www.jsc.nasa.gov/bu2/inflate.html
Internation Small Satellite Link Resources	140.116.200.187/guan/homepage/smatlink.htm
International Association for the Astronomical Arts	www.novaspace.com/IAAA/IAAA.shtml
ISSO New Space Network	www.isso.org/home-hi.html

Jane's Information Group	www.janes.com/janes.html
Jonathan's Space Report	hea-www.harvard.edu/QEDT/jcm/space/jsr/jsr.html
Journal of Small Satellite Engineering	www.ee.surrey.ac.uk/EE/CSER/UOSAT/IJSSE/ijsse.html
Keith Stein's Community, Air & Space Report	www.newspace.com/publications/casr/home.html
Launchspace Inc.	www.newspace.com/home.html
NASA Export Control Program	www.hq.nasa.gov/office/codei/nasaecp/TOCNASAECP.html
NASA Galaxie—Library Search System	nasagalaxie.larc.nasa.gov/altlogin2.html
NASA Historical Archive @ KSC	www.ksc.nasa.gov/history/history.html
NASA Information Services via World Wide Web	www.gsfc.nasa.gov/NASA_homepage.html
NASA Master Directory	nssdc.gsfc.nasa.gov/nmd/nmd.html
NASA RECON Report Server	www.sti.nasa.gov/RECONselect.html
NASA Scientific and Technical Information Facility (STIF)	www.casi.sti.nasa.gov/
NASA Scientific and Technical Information Server	www.sti.nasa.gov/
NASA Space Shuttle Web Archives	shuttle.nasa.gov/archives.html
NASA Spacelink @ HQ	spacelink.msfc.nasa.gov/
NASA Spacelink—An Aeronautics & Space Resource for Educators	spacelink.nasa.gov/.index.html
NASA Tech Briefs	www.nasatech.com/
NASA Technical Report Server (NTRS)	techreports.larc.nasa.gov/cgi-bin/NTRS
NASA Thesaurus	www.sti.nasa.gov/nasa-thesaurus.html
National AirSpace Information	www.nasi.hq.faa.gov/
National Science Foundation	www.nsf.gov/
National Space Science Data Center	nssdc.gsfc.nasa.gov/nssdc/nssdc_home.html
National Tech Transfer Center (NTTC) Gateway	iridium.nttc.edu/searching.html
New Space Network	www.newspace.com/
NOVAGRAPHICS SPACE ART GALLERY	www.novaspace.com/
NRC Homepage	www.nas.edu/
NSSDC's CDF Homepage	nssdc.gsfc.nasa.gov/cdf/cdf_home.html
Report from the First Origins Technology Workshop	poe.ipac.caltech.edu:8080/origins/workshop/origins_1.html
Satellite Engineer Online Magazine	www.satengineer.com/
Satellite Journal International	www.nmia.com/~roberts/sji/sj300.html
Satellite Times	www.grove.net/hmpgst.html
Small Satellite Technology Pages	www.gsfc.nasa.gov/gsfc/small_sat/small_sat.html
Small Spacecraft Missions Symposium	envnet.gsfc.nasa.gov:80/ssm/
Smithsonian Astrophysical Observatory	cfa-www.harvard.edu/sao-home.html
Space Access Society	www.space-access.org/
Space Calendar @ JPL	newproducts.jpl.nasa.gov/calendar/
Space Frontier Foundation	www.space-frontier.org/
Space News	www.spacenews.com/
Space Policy/Elsevier	www.elsevier.com/
Space Science Institute	www-ssi.colorado.edu/1.html
Space Studies Board	www.nas.edu/ssb/ssb.html

SpaceViews Home Page	www.seds.org/spaceviews/
Via Satellite Online	www.phillips.com/ViaOnline/
Visions Of Space	members.aol.com/space7/space7.html
World Spaceflight News	members.aol.com/wsnspace/index.htm

SCIENCE DATA ARCHIVES

Astrophysics Servers

GGS Mission Operations	ggsfot.gsfc.nasa.gov/
High Energy Astrophysics Science Archive	legacy.gsfc.nasa.gov/
NASA's Astrophysics Data System	http//adswww.harvard.edu/
Space Physics Satellite Data Base— NSSDC	nssdc.gsfc.nasa.gov/space/space_physics_home.html
Space Telescope Electronic Information Service	www.stsci.edu//

Earth Science Data Servers

CPSR's Home Page	snyside.sunnyside.com/dox/home.html#hot
Declassified Satellite Photographs	edcwww.cr.usgs.gov/dclass/dclass.html
EESC—Home Page	terrassa.pnl.gov:2080/
ERIC/CSMEE Home Page	gopher.ericse.ohio-state.edu/
EROS—Home Page	edcwww.cr.usgs.gov/eros-home.html
ESDIS Acronym List Apr 1997	gila.gsfc.nasa.gov/acronyms.html
ESOC External Home Page	www.esoc.esa.de/external/mso/
GLIS—Home Page	sun1.cr.usgs.gov/glis/glis.html
Global Change Master Directory	gcmd.gsfc.nasa.gov/
JPL PO.DAAC Home Page	podaac-www.jpl.nasa.gov/
NASA EPIMS-Web	epims.gsfc.nasa.gov/epims_pub/
NOAA—Home Page	www.noaa.gov/
NOAA/CDC Satellite Climate Research	www.cdc.noaa.gov/~climsat/
NOAA's Geostationary Satellite Browse Server	goeshp.wwb.noaa.gov/
OCEAN IMAGING—Satellite and Environmental Services	www.oceani.com/
Office of Research and Applications Home Page	orbit-net.nesdis.noaa.gov/ora/

Planetary Data Servers

Center for Mars Exploration Home Page	cmex-www.arc.nasa.gov/
Lunar and Planetary Institute	cass.jsc.nasa.gov/lpi.html
Mars Sample Return: Issues and Recommendations	www.nas.edu/ssb/mrsrexec.html
Planetary Data Center	cdwings.jpl.nasa.gov/PDS/
Planetary Science Institute	www.psi.edu/
Planetary Spacecraft Data Base— NSSDC	nssdc.gsfc.nasa.gov/planetary/planetary_home.html
SEDS Internet Headquarters	seds.lpl.arizona.edu/
Smithsonian Astrophysical Observatory	cfa-www.harvard.edu/sao-home.html
Solar System	lablinks.com/sumeria/chapiii.html
Solar System @ Los Alamos	www.c3.lanl.gov/~cjhamil/SolarSystem/solarsystem.html
Solar System on the Web	www.anu.edu.au/Physics/ssweb/Welcome.html
Views Of The Solar System	bang.lanl.gov/solarsys/

Adolphsen, John, "The Cosmic Ray Upset Experiment (CRUX)," presented at the Radiation and Their Effects on Components and Systems Symposium, Arcachon, France, November 1995.

Alexander, A., *The Cost and Benefit of Reliability in Military Equipment,* Santa Monica, Calif.: RAND, P-7515, December 1988.

Alfriend, K., et al., "Attitude and Orbit Determination of a Tethered Satellite System," presented at the AAS/AIAA Astrodynamics Specialist Conference, Halifax, Nova Scotia, August 1995.

Aljabri, A., "Autonomous Serendipitous Science Acquisition for Planets," presented at the 34th AIAA Aerospace Sciences Meeting and Exhibit, Pasadena, Calif.: NASA Jet Propulsion Laboratory, January 17, 1996.

Alkalai, Leon, et al., "Summary of Results from the Focus Groups," presented at the Workshop on Remote Exploration and Experimentation, Pasadena, Calif.: NASA Jet Propulsion Laboratory, August 29, 1995.

Arno, Roger, "Cheaper Science Missions: Reality, Hyperbole, or Delusion? Strictly Personal and Unofficial Observations," Moffet Field, Calif.: NASA Ames Research Center, October 1996.

Ayers, F., "The Management of Technological Risk," *Research Management,* November 1977.

Ballistic Missile Defense Organization, *Cost Risk Analysis of the Ballistic Missile Defense (BMD) System,* Washington, D.C., TR-9042-2, Rev. 2, February 1996.

Baluck, Michael J., et al., *Fundamentals of Plastic Encapsulated Microcircuits (PEMs) for Space Applications,* Greenbelt, Md.: NASA Goddard Space Flight Center, February 1995.

Barela, P., S. Bolin, S. Cornford, N. Elgabalawi, T. Gindorf, M. Krasich, and L. Wright, *μ-S/C Mission Assurance: Assessment Team Final Report,* Pasadena, Calif.: NASA Jet Propulsion Laboratory, October 17, 1994a.

_____, *μ-S/C Mission Assurance: RTOP Program Plan*, Pasadena, Calif.: NASA Jet Propulsion Laboratory, December 1994b.

Bearden, D., *Small Spacecraft Subsystem-Level Cost Model—Version 2.0: Final Report*, Los Angeles, Calif.: The Aerospace Corporation, MIPR-C-32034-E, November 4, 1996.

Beauchamp, Patricia M., "Miniature Integrated Camera Spectrometer (MICAS)," briefing, Pasadena, Calif.: NASA Jet Propulsion Laboratory, January 10, 1997.

Bedingfield, K., et al., *Spacecraft System Failures and Anomalies Attributed to the Natural Space Environment*, Huntsville, Ala.: NASA Marshall Space Flight Center, NASA RP-1390, August 1996.

Bowers, A., and J. Dertouzos, *Essays in the Economics of Procurement*, Santa Monica, Calif.: RAND, MR-462-OSD, 1994.

Brown, A., *NASA Unmanned Flight Anomaly Report: Analysis of Uplink/ Downlink Anomalies on Six JPL Spacecraft*, Pasadena, Calif.: NASA Jet Propulsion Laboratory, JPL-D-11383, September 1994.

Brown, A., et al., *JPL Common Threads Workshop: Summary Report*, Pasadena, Calif.: NASA Jet Propulsion Laboratory, JPL-D-13776, July 1996.

Brown, K., "GSFC Integrated Mission Design Center (IMDC)," briefing to industry, Greenbelt, Md.: NASA Goddard Space Flight Center, February 26, 1997.

_____, "Rapid Spacecraft Procurement: Briefing to Industry's Potential Bidders," Greenbelt, Md.: NASA Goddard Space Flight Center, Jaury 26, 1997.

Bruner, M., "Technology Issues for Small Scientific Satellites," presented at the First Goddard Workshop on Small Satellites, February 1995.

Brunschwyler, J., "A Non-Standard Production Approach to Satellite Constellations," presented at the 10th Annual AIAA/Utah State University Conference on Small Satellites, Logan, Utah, September 1996.

Brutocao, J., *Known Good Die: Facilitating Multi-Chip Modules*, Pasadena, Calif.: NASA Jet Propulsion Laboratory, April 29, 1994.

_____, *How Microelectronic Test Structures Impact Space Applications*, Pasadena, Calif.: NASA Jet Propulsion Laboratory, January 1, 1995.

Butler, Madeline, et al., "Renaissance: A Revolutionary Approach for Providing Low-Cost Ground Data Systems," presented at the 4th International Symposium on Space Mission Operations and Ground Data Systems, Munich, Germany, September 17, 1996.

Cameron, G., et al., "System Engineering of Cost Efficient Operations," presented at the 4th International Symposium on Space Mission Operations and Ground Data Systems, Munich, Germany, September 17, 1996.

Casani, E. Kane, "The New Millennium Program: Management Challenges in the 21st Century," presented at the IEEE Aerospace Applications Conference, Aspen, Colo., February 1996.

_____, "The New Millennium Program: Technology Development for the 21st Century," presented at the 34th AIAA Aerospace Sciences Meeting and Exhibit, Reno, Nev., January 15, 1996.

Casani, E., et al., "The Flight System Testbed," *SPIE Space Instrumentation and Dual-Use Technologies Proceedings*, Vol. 2214, June 1994.

Chilcot, K., *Application of Risk Assessment Techniques in Optimizing Future Space Missions*, AIAA Space Programs and Technologies Conference, AIAA-88-3509, Houston, Tex., June 21, 1988.

Chow, B., *Emerging National Space Launch Programs*, Santa Monica, Calif.: RAND, R-4179-USDP, 1993.

Coppola, A., "Status of Reliability Technology," *IEEE Reliability Society Newsletter*, April 1995.

Cordell, Bruce, "Forecasting the Next Major Thrust into Space," *Space Policy*, Vol. 12, No. 1, February 1996, pp. 45–57.

Cornford, Steven, "Defect Detection and Prevention in the Risk Management Process," NASA Risk Management Course briefing, Pasadena, Calif.: NASA Jet Propulsion Laboratory, November 14, 1996a.

_____, "JPL Test Effectiveness Program," briefing to the NASA Office of Safety and Mission Assurance, Pasadena, Calif.: NASA Jet Propulsion Laboratory, December 11, 1996b.

Creech, S., NASA Marshall Space Flight Center, "New Ways of Doing Business Review," presented at the AIAA Management Council Meeting, Arlington, Va., May 1996.

Creedon, J., et al., *NASA Technology Transfer: Report of the NASA Technology Transfer Team (Creedon Report)*, Hampton, Va.: NASA Langley Research Center, December 21, 1992.

David, Leonard, "Observing a Failure," *Aerospace America*, June 1994, pp. 28–33.

_____, "Severity of Orbital Debris Questioned," *Space News*, January 19, 1997a.

_____, "New Technology Proves Difficult for Deep Space 1," *Space News*, September 22, 1997b.

_____, "Incredible Shrinking Spacecraft," *Aerospace America*, January 1996.

Defense Systems Management College, *Risk Management Concepts and Guidance*, Fort Belvoir, Va., ADA-214-342, 1997.

_____, *Systems Engineering Management (SEM) Guide*, 2nd ed., Fort Belvoir, Va., ADA-223-168, 1990.

"Demise of NASA Advanced Technology Unit Viewed with Alarm," *Space Business News*, August 7, 1996.

Dole, S. H., et al., *Methodologies for Analyzing the Comparative Effectiveness and Cost of Alternative Space Plans*, Vol. 1: *Summary*, Santa Monica, Calif.: RAND, RM-5656-NASA, 1968.

EIA/IEEE/ISHM Joint Meeting on Electronic Components for the Commercialization of Military and Space Sytems, Workshop Notes, San Diego, Calif.: February 3, 1997.

Figueroa, O., and Gilberto Colón, "SAMPEX," in Wertz and Larson (1996), pp. 397–409.

Fleeter, Rick, "Management of Small Satellite Programs—Lessons Learned," presented at the IAF/COSPAR World Space Congress, Washington, D.C., September 2, 1992.

_____, *Micro Space Craft*, Reston, Va.: Edge City Press, 1995.

_____, "Mr. Murphy on Small Spacecraft and Rocket Reliability," *Launchspace Magazine*, March 1997a, p. 14.

_____, "Component Selection for Low-Cost Spacecraft: Parts is Parts?" *New Space Newsletter*, Spring 1997b.

Forman, Brenda, "What Am I Bid: Market-Based Science Planning for Planetary Spacecraft," *SpaceWatch*, January 1997.

Garrison, A., *High Density Spaceflight Application of PEMS with Paralyne Conformal Coating*, Greenbelt, Md.: NASA Goddard Space Flight Center, 1996.

Gavit, Sarah A., "Mars Microprobe Project," briefing presented to L. Sarsfield, NASA New Millennium Program, January 10, 1997.

Gibbons, John H., *Remarks of the President's Science Advisor, Dr. John H. Gibbons, at the Wernher von Braun Lecture*, National Air and Space Museum, Washington, D.C., March 22, 1995.

Gilman, D., *Report from the NASA Workshop on Quality and Mission Life Cycle Costs*, NASA of Office for Safety and Mission Assurance, December 1993.

Gindorf, Tom, *NASA Approach to Spacecraft Product Assurance: Typical JPL Program*, Pasadena, Calif.: NASA Jet Propulsion Laboratory, April 24, 1995.

Gindorf, Tom, et al., "Space-Hardware Design for Long Life with High Reliability," *Proceedings of the IEEE Annual Reliability and Mainatainabiltiy Symposium*, Anaheim, Calif., January 1994a.

_____, *Micro-Spacecraft Mission Assurance: Assessment Team Final Report*, Pasadena, Calif.: NASA Jet Propulsion Laboratory, October 17, 1994b.

_____, "Environmental Test Effectiveness," presented to the IES 15th Aerospace Testing Seminar, Los Angeles, Calif.: October 1994c.

Gipson, Melinda, and Richard Buenneke, *Microspace Renaissance*, Arlington, Va.: Pasha Publications, 1992.

Glicksman, R., *Estimating Collision Probability for Coincident Satellite Constellations*, Boulder, Colo.: Space Analytics Associates, Inc., AIAA-96-3635-CP, 1996.

Gonzalez, C., *Correlation of the Magellan Flight PFR History with Ground-Test Results*, Pasadena, Calif.: NASA Jet Propulsion Laboratory, JPL-D-12771, January 1996a.

_____, *In-Flight Parts-Related Problem/Failures*, Pasadena, Calif.: NASA Jet Propulsion Laboratory, JPL D-13482, September 1996b.

Graves, H., et al., *Simulation Based Design Phase 2*, Lockheed Martin Missiles and Space Company, February 13, 1997.

Greenfield, M., *Risk as a Resource*, NASA OSMA, June 24, 1997.

Gregory, Bill, "The DBOF's Demise: Depot and Shipyard Costs Prove Resistant to Innovative Pentagon Financial Accounting Practices," *Armed Force Journal International*, June 1997.

Groszyk, W., *Primer on Performance Measurement*, Office of Management and Budget, Washington, D.C., 1995.

GSFC—*See* National Aeronautics and Space Administration, Goddard Space Flight Center.

Hamaker, J., *But What Will it Cost?—The Evolution of NASA Cost Estimating*, NASA Marshall Space Flight Center, 1991.

Hammer, M., *Beyond Reengineering*, New York: HarperBusiness, 1996.

Harmon, B. R., *Assessing Acquisition Schedules for Unmanned Spacecraft*, Alexandria, Va.: Institute for Defense Analysis, April 1993.

Hartley, Jonathan, et al., "Automation of Satellite Operations: Experiences and Future Directions at NASA GSFC," presented at the 4th International Symposium on Space Mission Operations and Ground Data Systems, Munich, Germany, September 17, 1996.

Hecht, H., "Reliability During Space Mission Concept Exploration," in Larson and J. Wertz (1992), pp. 700–714.

_____, "Reliability Considerations," in Wertz and Larson (1996), pp. 285–302.

Hecht, M., et al., "Causes and Effects of Spacecraft Failures," *Quality and Reliability Engineering International*, Vol. 4, 1988, pp. 11–19.

Hecht, M., C. Kukkonen, and S. Venneri, "In-Situ Instruments and Micro-Electromechanical Systems (MEMS) for Planetary Science," paper presented at the 47th International Astronautical Congress, Beijing, China, October 7–11, 1996, Paris, France: International Astronautical Federation, IAF-96-A.2.0.7, 1996.

Hemmings, J., "NEAR Costing as a Template for Future Small Spacecraft Missions," presented at 10th Annual AIAA/Utah State University Conference on Small Satellites, Logan, Utah, September 1996.

Hunsucker, J. L., *Integrated Risk Management, Final Report of the 1993 NASA Summer Faculty Fellowship Program*, Houston, TX: NASA Johnson Space Center, N94-25361, August 1993.

Huntress, W., "dear colleague" letter, NASA Office of Space Science, February 1995a.

_____, Associate Administrator, NASA Office of Space Science, "Center Roles in Space Science," dear colleague letter, May 23, 1995b.

Intel Corporation, *Intel's Quality System*, 1996.

International Standards Organization, "Space Systems Programme Management," Committee Draft, ISO-CD-14300, Geneva, Switzerland, August 11, 1996.

Isakowitz, S., *International Reference Guide to Space Launch Systems*, 2nd ed., 1994.

Jarosz, Mark, et al., "Integration and Test of X-Ray Timing Explorer Spacecraft Attitude Control System," presented at the 18th Annual AAS Guidance and Control Conference, Keystone, Colo., AAS-95-025, February 2, 1995.

Jordan, J., et al., "How Burn-In Can Reduce Quality and Reliability," *Proceedings of the Institute of Environmental Sciences*, 1996, p. 18.

JPL—*See* National Aeronautics and Space Administration, Jet Propulsion Laboratory.

Kallender, Paul, "Failure of ADEOS Fuels Debate of Size of Satellites," *Space News*, July 7, 1997.

Katz, A., "Kelly Johnson on the Weapons Acquisition Process," Santa Monica, Calif.: RAND, PR-20444, June 1970.

Kelada, Joseph, *Integrating Reengineering with Total Quality*, Milwaukee, Wisc.: American Society of Quality Control Press, 1996.

Kerslake, William R., "Development and Flight History of the SERT II Spacecraft," AIAA/SAE/ASME/ASEE 28th Joint Propulsion Conference, Nashville, Tenn., AIAA-92-3516, July 6, 1996.

Kicza, M., et al., *NASA Draft Response to RAND Interim Report on Federal Investments in Small Spacecraft*, April 1997.

Kostoff, R., *Research Program Peer Review: Principles, Practices, Protocols*, Arlington, Va.: Office of Naval Research, August 1996.

_____, *The Handbook of Research Impact Assessment*, 7th ed., Arlington, Va.: Office of Naval Research, ADA-296021, May 1997.

_____, "Peer Review: The Appropriate GPRA Metric for Research," *Science*, August 1, 1997, p. 651.

Krasich, M., Jet Propulsion Laboratory, "Reliability Prediction Using Flight Experience: Weibull Adjusted Probability of Survival Method," University of Maryland Lecture on Reliability, April 1995.

Krasner, Sanford M., and Douglas E. Bernard, "Integrating Autonomy Technologies into an Embedded Spacecraft System Flight Software System Engineering for a New Millennium," presented at 10th Annual AIAA/Utah State University Conference on Small Satellites, Logan, Utah, September 1996.

LaBel, Kenneth, et al., *Commercial Microelectronics for Applications in the Satellite Radiation Environment*, Greenbelt, Md.: NASA Goddard Space Flight Center, 1996.

Larson, W., "Process Changes to Reduce Cost," in Wertz and Larson (1996), pp. 17–53.

Larson, W., and J. Wertz, eds., *Space Mission Analysis and Design*, 2nd ed., Torrance, Calif.: Microcosm, Inc., and Dordrecht, The Netherlands: Kluwer Academic Publishers, 1992.

Lau, Kenneth, Stephen Lichten, Lawrence Young, and Bruce Haines, "An Innovative Deep Space Application of GPS Technology for Formation Flying Spacecraft," presented at the American Institute of Aeronautics and Astronautics Guidance, Navigation, and Control Conference, San Diego, Calif., AIAA 96-3819, July 1996.

Lauriente, M., et al., "Spacecraft Anomalies due to Radiation Environment in Space," presented at the 2nd International Workshop on Radiation Effects of Semiconductor Devices for Space Applications, Tokyo, Japan, March 21, 1996.

Lawler, Andrew, "Small Missions Lift Planetary Science," *Science*, September 12, 1997.

Lehman, D., "Into the New Millennium: Deep Space One," NASA GSFC Small Spacecraft Mission Symposium, October 1996.

Lucau, P., "Better, Faster and Cheaper Versus Mission Success," paper presented at the 47th International Astronautical Congress, Beijing, China, October 7–11, 1996, Paris, France: International Astronautical Federation, IAA-96-IAA.6.1.03, 1996.

Man, Guy K., "Why Do We Need to Flight Validate the Remote Agent?" briefing, NASA New Millennium Program, January 10, 1997.

Maurer, R., "The NEAR Discovery Mission: Lessons Learned," presented at the 10th Annual AIAA/Utah State University Conference on Small Satellites, Logan, Utah, September 1996.

McCauley, Molly, *Measuring Results from NASA Funding of Technology Intended for Commercial Markets: An Overview*, Working Paper, Washington, D.C.: Resources for the Future, November 1996.

Meyer, D., *Design and Manufacture of Space-Qualified MEMS Components*, Los Angeles, Calif.: The Aerospace Corporation, June 1995.

Moore, N.R., et al., *An Improved Approach for Flight Readiness Certification: Methodology for Failure Risk Assessment and Application Examples*, Pasadena, Calif.: NASA Jet Propulsion Laboratory, JPL-92-15, June 1992.

Muirhead, B., *Mars Pathfinder Flight Systems Design and Implementation*, Pasadena, Calif.: NASA Jet Propulsion Laboratory, August 1996.

Mulville, D., *NASA Draft Technology Implementation Plan and Technology Leadership Council Charter*, September 4, 1996.

Musser, George, "Faster, Better, Cheaper, How? An Interview with Domenick J. Tenerelli," *Mercury Magazine*, Vol. 24, No. 4, July 1995.

NASA—*See* National Aeronautics and Space Administration.

National Aeronautics and Space Administration, *Reliability Preferred Practices for Design and Test*, Washington, D.C., NASA-TM-4322, n.d.

_____, *Management of Major System Programs and Projects*, Washington, D.C., NHB 7120.5A, 1996a.

_____, "Understanding Our Changing Planet: NASA's Mission to Planet Earth," briefing, Washington, D.C., April 1996b.

_____, *NASA Strategic Management Handbook*, Washington, D.C., October 1996c.

National Aeronautics and Space Administration, Advisory Committee on the Future of the U.S. Space Program (Augustine Committee), *Report of the Advisory Committee on the Future of the U.S. Space Program*, Washington, D.C.: U.S. Government Printing Office, December 1990.

National Aeronautics and Space Administration, Advisory Council, Space and Earth Science Advisory Committee, *The Crisis in Space and Earth Science*, Washington, D.C., November 1986.

National Aeronautics and Space Administration, Advisory Council, *NASA Federal Laboratory Review*, Washington, D.C., February 1995.

National Aeronautics and Space Administration, Advisory Council, Space Science Advisory Committee, Minutes, May 15, 1996.

National Aeronautics and Space Administration, Goddard Space Flight Center, Engineering Directorate, *Charter of the Space Technology Division*, Greenbelt, Md., 1996a.

_____, *Earth Radiation Budget Satellite (ERBS) On-Orbit Reliability Investigation Report*, Greenbelt, Md., October 1996b.

_____, *Spartan 207 Inflatable Antenna Experiment: Preliminary Mission Report*, Greenbelt, Md.: Spartan Project Office, February 1997.

_____, *Qualification of Non-Standard EEE Parts in Spaceflight Applications*, Greenbelt, Md., NASA Preferred Reliability Practice PT-TE-1418, n.d., a.

_____, *Spacecraft Orbital Anomaly Reporting (SOAR) System*, Greenbelt, Md., Reliability Practice No. PD-ED-1232, n.d., b.

National Aeronautics and Space Administration, Headquarters, *Risk Management Policy for Manned Flight Programs*, Washington, D.C., NMI 8070.4, June 1994.

_____, *Overview of the New Millennium Program*, Attachment 1, NMP Integrated Product Development Team Solicitation, Washington, D.C., May 1995.

_____, *Budget Estimates: Fiscal Year 1998*, 1997.

_____, *NASA Participation in the Development and Use of Voluntary Standards*, NMI 8070.2B [expired], Washington, D.C., September 1991.

National Aeronautics and Space Administration, Headquarters, Office of Policy and Plans, *Strategic Planning and Strategic Management at NASA*, Washington, D.C., June 1996.

National Aeronautics and Space Administration, Jet Propulsion Laboratory, *Analysis of Uplink/Downlink Anomalies on Six JPL Spacecraft*, NASA Unmanned Flight Anomaly Report, Pasadena, Calif., JPL D-11383, September 1994.

_____, *New Millennium Program Technology Selection Process Plan*, Pasadena, Calif., JPL D-13361 Rev. B, May 10, 1996a.

_____, "Instrument Technologies and Architectures Technology Roadmap," Pasadena, Calif.: New Millennium Program, February 199b6.

_____, New Millennium Program, "Cost Projection—NMP Deep Space One," Pasadena, Calif., June 1996c.

_____, *PDC Facility History and Forecast*, Pasadena, Calif., n.d.

National Aeronautics and Space Administration, Johnson Space Center, *Designing for Dormant Reliability*, Greenbelt, Md., Reliability Guideline No. GD-ED-2207, n.d.

National Aeronautics and Space Administration, Lewis Research Center, *Mission Statement of the Photovoltaic Branch*, Cleveland, Ohio, 1995.

National Aeronautics and Space Administration, Office of the Administrator, *Customer Satisfaction Report*, Washington, D.C., 1995a.

_____, *Fiscal Year 1995 Accountability Report*, Washington, D.C., 1995b.

National Aeronautics and Space Administration, Office of Aeronautics and Space Technology, *NASA Integrated Technology Plan*, Washington, D.C., 1991.

_____, *NASA Strategic Plan*, Working Draft 4A, Washington, D.C., April 10, 1997.

National Aeronautics and Space Administration, Office of the Chief Financial Officer, *Full-Cost Accounting*, Washington, D.C., 1997a.

_____, *NASA Full-Cost Initiative—Agencywide Test Guide*, draft, Washington, D.C., March 1997b.

National Aeronautics and Space Administration, Office of the Mission to Planet Earth, Announcement of Opportunity, Washington, D.C., July 1996.

National Aeronautics and Space Administration, Office of Safety and Mission Assurance, "OMSA—Long-Range Planning for Space Technology Development," presented at the AIAA Space Programs and Technologies Conference, Huntsville, Ala., March 24, 1992.

_____ *Safety and Mission Assurance Strategic Plan*, Washington, D.C., April 19, 1996.

National Aeronautics and Space Administration, Office of Space Access and Technology, "Report from the First Origins Technology Workshop," Washington, D.C.: June 4, 1996.

National Aeronautics and Space Administration, Office of Space Science, *Integrated Technology Strategy*, Washington, D.C., 1995a.

_____, *Space Science for the 21st Century*, Washington, D.C., August 1995b.

_____, *Space Technology Data Base, Advanced Battery Requirement*, Washington, D.C., June 1996a.

_____, *Science in Air and Space: NASA's Science Policy Guide*, Washington, D.C., July 1996b.

National Aeronautics and Space Administration, Office of Space Science, Astrophysics Division, Astrophysics Subcommittee, Minutes, Washington, D.C., September 1994.

National Aeronautics and Space Administration, Office of Space Science and Applications, *A Strategy for Leadership in Space Through Excellence in Space Science and Applications: An OSSA Strategic Plan*, Washington, D.C., 1988.

National Aeronautics and Space Administration, Office of Space Science, Solar System Exploration Division, Solar System Exploration Subcommittee, *Solar System Exploration, 1995–2000*, Washington, D.C., September 1994.

National Aeronautics and Space Administration, Office of Space Science, Solar Physics Division, "FY '96 Solar Physics SR&T AO Announcement," Washington, D.C., May 1996.

National Aeronautics and Space Administration, Space Systems and Technology Advisory Committee, *Advanced Technology for America's Future in Space*, Washington, D.C., 1991.

National Research Council, *Space Technology to Meet Future Needs*, Washington, D.C., 1987.

National Research Council, *A Science Strategy for Space Physics*, Joint Report of the Committee on Solar and Space Physics and the Committee on Solar-Terrestrial Research, Washington, D.C., 1994.

National Research Council, "Presentation to the Aeronautics and Space Engineering Board of the NRC, Workshop on Reducing the Costs of Space Science Research Missions," Group One Team Report, Irvine, Calif., October 1996.

National Research Council, Aeronautics and Space Engineering Board, *Technology for Small Spacecraft*, Washington, D.C., 1994.

National Research Council, Aeronautics and Space Engineering Board and Space Studies Board, Joint Committee on Technology for Space Science and Applications, *Reducing the Costs of Space Science Research Missions: Proceedings of a Workshop*, Washington, D.C., 1997.

National Research Council, Committee on Space Science Technology Planning, *Improving NASA's Technology for Space Science*, Washington, D.C., 1993.

National Research Council, Space Studies Board, *Assessment of Recent Changes in the Explorer Program*, March 1997.

National Research Council, Space Studies Board, Commission on Physical Sciences, Mathematics, and Applications, "Letter Report on the Optimum Phasing for SIRTF," Washington, D.C.: February 2, 1996b.

National Research Council, Space Studies Board, Committee on Planetary and Lunar Exploration, *The Role of Small Missions in Planetary and Lunar Exploration*, Washington, D.C., 1995.

National Research Council, Space Studies Board, Committee on Planetary and Lunar Exploration, *Review of NASA's Planned Mars Program*, Washington, D.C., 1996a.

National Science and Technology Council, *Federal R&D Investment Strategies for Satellite Technologies*, draft, Washington, D.C., March 1996.

Newell, Homer, *Beyond the Atmosphere*, Washington, D.C., National Aeronautics and Space Administration, NASA SP-4211, 1980.

Noor, Ahmed K., et al., "A Virtual Design Environment for Intelligent Design," *Aerospace America*, April 1997.

NRC—*See* National Research Council.

NRC SSB—*See* National Research Council, Space Studies Board.

OAST—*See* National Aeronautics and Space Administration, Office of Aeronautics and Space Technology.

Oberhettinger, David, *Investigation of Mechanical Anomalies Affecting Interplanetary Spacecraft*, NASA Unmanned Flight Anomaly Report, Pasadena, Calif.: NASA Jet Propulsion Laboratory, JPL-D-11951, September 1994.

_____, *Investigation of Environmentally Induced Anomalies Aboard JPL Spacecraft*, NASA Unmanned Flight Anomaly Report, Pasadena, Calif.: NASA Jet Propulsion Laboratory, JPL-D-12546, June 1995a.

_____, *Product Assurance Measures Applicable to Prevention of JPL In-Flight Anomalies*, Pasadena, Calif.: NASA Jet Propulsion Laboratory, JPL-D-12840, September 1995b.

Office of Management and Budget, confidential correspondence, April 1996.

OMB—*See* Office of Management and Budget.

OSS—*See* National Aeronautics and Space Administration, Office of Space Science.

OSSA—*See* National Aeronautics and Space Administration, Office of Space Science and Applications.

Pace, Scott, Gerald Frost, Irving Lachow, David Frelinger, Donna Fossum, Donald K. Wassem, and Monica Pinto, *The Global Positioning System: Assessing National Policies*, Santa Monica, Calif.: RAND, MR-614-OSTP, 1995.

Pecht, M., "Issues Affecting Early Affordable Access to Leading Electronics Technologies by the U.S. Military and Government," *Circuit World*, Vol. 22, No. 2, 1996a.

_____, "Why Traditional Reliability Prediction Models Do Not Work," *Electronics Cooling*, January 1996b.

_____, "Plastic Encapsulated Microcircuits," *Engineering Magazine*, September 1996c.

Pecht, M., et al., "Are Components Still the Major Problem: A Review of Electronic System and Device Field Failure Returns," *IEEE Transactions on Components, Hybrids, and Manufacturing Technology*, Vol. 15, No. 6, December 1992, p. 1163.

_____, *Plastic Encapsulated Microelectronics*, New York: Wiley & Sons, 1994.

Perry, William J., Secretary of Defense, *Specifications & Standards—A New Way of Doing Business*, June 29, 1994.

Plum, Paul, "New Blueprint for ESS," *Quality Magazine*, November 1990.

Priore, N., and Farrell, J., *Plastic Microcircuit Packages: A Technology Review*, Rome, N.Y.: Reliability Analysis Center, CRT-PEM, March 1992.

Remez, J., et al., *Orbital Anomalies in Goddard Spacecraft for Calendar Year 1995*, Greenbelt, Md.: NASA Goddard Space Flight Center, 1996.

Research Performance Measures Round Table, "Developing and Presenting Performance Measures for Research Programs," Washington, D.C., August 1995.

Rhea, J., "The Challenge of Space on the New COTS Frontier," *Military and Aerospace Electronics*, May 1997.

Richardson, Gareth, Nizar Sultan, and Alfred Ng, "Parametric Design Curves for Payload Power and Mass Capabilities of Non-Geo Smallsats

Buses/Launchers," presented at 10th Annual AIAA/Utah State University Conference on Small Satellites, Logan, Utah, September 1996.

Ridenoure, R., NASA Jet Propulsion Laboratory, "Key Architectural Issues and Trade-Offs for the New Millennium Advanced Technology-Validation Models," presented at the AIAA 34th Aerospace Sciences Meeting, Reno, Nev., January 1996.

Robinson, P., *Spacecraft Environmental Anomalies Handbook*, Air Force Systems Command, Geophysics Center, TR-89-0222, 1989.

Rosen, Robert, et al., "Advanced Technologies to Support Earth Orbiting Systems," presented at the 43rd Annual IAF Congress, Washington, D.C.: August 28, 1992.

Ryschkewitsch, M., *Issues in Spacecraft Technology for Small Spacecraft*, First Goddard Workshop on Small Satellites, February 1995.

Sander, M., "Develop New Products Project," briefing presented to Mr. Liam Sarsfield, Pasadena, Calif.: NASA Jet Propulsion Laboratory, January 10, 1997.

Schoonwinkel, A., G. W. Milne, S. Mostert, W. H. Steyn, and K. van der Westhuizen, "Pre-Flight Performance of SUNSAT, South Africa's First Remote Sensing and Packet Communications Microsatellite," presented at 10th Annual AIAA/Utah State University Conference on Small Satellites, Logan, Utah, September 1996.

Schultz, William, et al., *Reliability Considerations for Using Plastic-Encapsulated Microcircuits in Military Applications*, Melbourne, Fla: Harris Semiconductor, September 8, 1994.

Seidleck, Christina M., et al., *Single Event Effect Data Analysis of Multiple NASA Spacecraft and Experiments: Implications to Spacecraft Electrical Designs*, Lanham, Md.: Hughes STX, 1995.

Selding, Peter de, "European, Russian Satellites Have Close Call in Orbit," *Space News*, August 3, 1997.

Semiconductor Industry Association, *World Semiconductor Trade Statistics*, San Jose, Calif., 1996.

Servais, G., "The Evolution of Plastic ICs and Their Reliability in Automotive Applications," briefing 1997. Reprinted in notes from Electronic Components for the Commercialization of Military and Space Systems, held in San Diego, Calif., on February 3–5, 1997.

SESAC—*See* NASA Advisory Council, Space and Earth Science Advisory Committee.

Siewert, S., Colorado Space Grant College, First GSFC Workshop on Autonomous "Lights Out" Operations, October 1996, pp. 3 and 26.

SMC—*See* U.S. Air Force Materiel Command, Space and Missile Systems Center.

Smith, David B., "Develop New Products (DNP) System Engineering," briefing, Pasadena, Calif.: NASA Jet Propulsion Laboratory, March 15, 1996.

_____, *Reengineering Space Projects*, Pasadena, Calif.: NASA Jet Propulsion Laboratory, January 1997.

SSG, Inc., "Advanced Silicon Carbide Optics and Structures as Applied to Integrated Sensors," briefing, Pasadena, Calif.: NMP Instruments IPDT, February 1996.

Stadterman, T., et al., *A Physics-of-Failure Approach to Accelerated Life Testing of Electronic Components*, U.S. Army Materiel Systems Analysis Activity (AMSAA), 1996.

Stern, Lawrence, et al., *An Assessment of Potential Markets for Small Satellites*, Fairfax, Va.: Virginia Center for Innovative Technology, November 1989.

Strope, D., X-*Ray Timing Explorer—A Standard of Excellence*, Columbia, Md.: COST, Inc. for the NASA Goddard Space Flight Center's Explorer Program, September 1996.

Tandler, J., "Automating the Operations of the Orbcomm Constellation," briefing to the first GSFC Workshop on Autonomous "Lights Out" Operations, Greenbelt, Md.: ORBCOMM Global, Limited Partnership, October 1996.

"TRW Refutes Skeptics on Long-Term Future," *Aviation Week & Space Technology*, January 16, 1995, p. 50.

Tyson, K. W., *A Perspective on Acquisition of NASA Space Systems*, Institute for Defense Analysis, December 1992.

Ulrich, P., "Advanced Technology and Mission Studies with OSS," briefing to the National Research Council Space Studies Board, March 4, 1997.

"U.S. Agencies Work to Form Next-Generation Satellites," *Space News*, February 3, 1997.

U.S. Air Force Materiel Command, Space and Missile Systems Center, "Management of Parts, Materials, and Processes," Los Angeles Air Force Base, Calif., SMC Regulation 800-34, Part 3, n.d.

U.S. General Accounting Office, *High Risk Series: NASA Contract Management*, Washington, D.C., GAO/HR-93-11, December 1992.

_____, *Causes and Impacts of Cutbacks to NASA's Outer Solar Systems Exploration Missions*, Washington, D.C., GAO-NSIAD-94-24, December 1993.

_____, *Management Reform: Implementation of the National Performance Review's Recommendations*, Washington, D.C., GAO-OCG-95-1, December 1994.

U.S. House of Representatives, Committee on Science and Technology, Subcommittee on Space Science and Applications, *United States Civilian Space Programs: 1958–1978*, Washington, D.C., January 1981.

U.S. House of Representatives, Committee on Science, Space and Technology Hearing Report, "NASA Procurement Reform," Hearing Vol. 108, March 1992.

U.S. Senate, Committee on Commerce, Science, and Transportation, Hearing on NASA and NSF Program Efficiencies, *Remarks of Thomas Schultz—Associate Director GAO*, Washington, D.C., June 24, 1997.

Watzin, J., "Evolution of SMEX," program briefing, Greenbelt, Md.: NASA Goddard Space Flight Center, March 2, 1995.

_____, *SMEX-Lite—NASA's Next Generation Small Explorer*, Greenbelt, Md.: NASA Goddard Space Flight Center, September 1996a.

_____, *SMEX Mission Perspectives*, J. Watzin, NASA Small Spacecraft Mission Symposium, October 1996b.

Weiss, Stanley, et al., "The Air Force and Industry Think Lean," *Aerospace America*, May 1996.

Wertz, James, "Radical Cost Reduction Methods," in Wertz and Larson (1996), pp. 34–53.

Wertz, James, and Simon Dawson, "What's the Price of Low Cost?" paper presented at the 10th Annual AIAA/USU Conference on Small Satellites, Torrance, Calif.: Microcosm, Inc., 1996.

Wertz, James, and Wiley Larson, *Reducing Space Mission Cost*, Torrance, Calif.: Microcosm, Inc., and Dordrecht, The Netherlands: Kluwer Academic Publishers, 1996.

Wessen, R., et al., "Market-Based Approaches to Managing Science Return from Planetary Missions," presented at the 4th International Symposium on Space Mission Operations and Ground Data Systems, Munich, Germany, September 17, 1996.

The White House, *Federal Procurement Reform*, Washington, D.C., Executive Order 12931, October 12, 1994.

_____, *The National Space Policy*, Washington, D.C., September 19, 1996.

The White House, Office of the Vice President, *Accompanying Report of the National Performance Review*, Washington, D.C., September 1993.

Wong, R., "Cost Modelling," in Larson and Wertz (1992), pp. 715–740.

Worden, Simon P., "Management on the Fast Track," *Aerospace America*, November 1994, pp. 30–33.